A PASSION
FOR PLANTS

FROM THE RAINFORESTS OF BRAZIL
TO KEW GARDENS

A PASSION FOR PLANTS

THE LIFE AND VISION OF
GHILLEAN PRANCE

Director of the Royal Botanic Gardens, Kew

BY **CLIVE LANGMEAD**

Royal Botanic Gardens, Kew
2001

Second edition 2001
(First edition published by Lion Publishing, 1995)

Cover design: Jeff Eden.
Page make-up: Media Resources, Information Services Department,
Royal Botanic Gardens, Kew

ISBN 1 900347 76 8

Printed in Great Britain by The Cromwell Press Limited

CONTENTS

1. THE DIRECTOR

The roar of heavy jets lumbering in low on the approach to London Heathrow airport drowns out the first quiet words of welcome offered me. I am with the director of the Royal Botanic Gardens, Kew, senior figure in the international botanical community.

He pauses uncertainly and fingers his greying beard for a moment, and hovers, apparently awkwardly, by the door of his office. It is a modest room strewn chaotically with a fascinating mix of letters, work files, slide boxes and tropical botanic memorabilia.

The noise fades. He holds out his hand once again and repeats his welcome, adding a personal apology for the thundering jet, as though the aircraft is a regrettable aspect of modern botany for which he is responsible.

Professor Ghillean Tolmie Prance MA, DPhil, DSc, FRS, FLS, FIBiol, FRGS, is a local expert on Heathrow landing patterns. If pressed he can describe precisely the routes the ponderous jumbos take as they cross his 'patch' to come into land. The patch in question is, of course, the three hundred fresh green acres of west London known to the public as 'Kew Gardens' but more properly called the Royal Botanic Gardens, Kew.

These gardens, together with an additional five hundred acres of rural estate at Wakehurst Place in Sussex, form the most extensive and influential botanic gardens in the world – which makes the director a man to be reckoned with when it comes to plants.

His conversation is reserved, somewhat clipped. Every statement is well considered. An ordered and scientific framework is placed upon our discussion, which for my purpose concerns the man himself. This presents something of a problem, for he rarely talks about himself – only about his work. In fact, he comments at the end of our morning that he has *never* spent so much time talking about himself, and he finds it exhausting. I suspect this is partly because his quite remarkable career defies most reasoned analysis, even when conducted by his own keen mind, and partly because his personality is so naturally and genuinely self-effacing that it is a real effort for him to concentrate upon it.

His statements in response to my questions are clear and pointed, and explanation is added only on request. He uses the words he needs, no more. He is, I discover later without surprise, a ruthless mover of business through committee (and there are dozens at Kew). Analysis of aircraft landing patterns is as natural to him as an assessment of the rate of visitor flow at Kew's Victoria Gate, deducing the function of unusual stamens on a new plant species, or listing the problems of crop recovery following drought in north-east Brazil.

It is all good grist to the mental mill, and it may one day come in useful. But in any case for him it is instinctive, fun to do and will stay carefully labelled in his mental attic until called for.

It is in this vitality of mind and in the gentle phrases, carefully and directly delivered, that the strength of the man shows. In fact he seems strangely vulnerable as he speaks, as though drawing only tentative conclusions. But the words are compelling and his observations demand consideration. After a few minutes it is all I can do to avoid putting 'sir' at the end of my questions. But I sense it would not be appreciated and perhaps even be offensive to such a courteous host.

The director is tall, though his body is spare, almost thin and lacking in physical presence. He seems to stoop, but it is an effect of the professorial style. He certainly looks the part. The beard, the slightly shabby suit, the shy fussy glance over the random piles of paper spilling across the office tables. At any moment I expect him to push aside a box of documents, discover a microscope and bend eagerly over it. You would not guess that this professor has led expeditions to the furthest reaches of the Amazon for thirty years, that he can strip down a Johnson outboard motor or a Land Rover in a jungle clearing, or ride a pack mule bareback for miles through dense rainforest.

There is a survivor's strength in the modest frame. Countless lacerations, infections, fevers and deprivations endured on tropical expeditions testify to that. Indeed from first impressions you might never guess a whole raft of things about Iain (the shortened version of Ghillean) Prance.

If I had thought I was dealing solely with a distinguished professor of plant taxonomy (the science of plant classification) I could not have been more wrong. It would be as if I had thought of Kew as simply a 'pleasant historic garden'. For in the garden and the man, there is a great deal more than there at first seems.

In fact they have a lot in common. Quiet direct style, high purpose and formidable achievement. Even as a scientist Iain's public profile is remarkable. He is a well-known speaker on botany, the environment, ecology, the rainforest, Kew itself and much else besides. He is wholly familiar with the microphone, lectern and broadcast studio. Just as Kew represents itself as a splendid, blossoming, organically-managed botanical garden cheerfully open to the public, so too 'Professor Ghillean Prance FRS' is the available public face of 'green' science. There is with them both a central mission to explain, inform and educate on plant matters.

Nonetheless he also stands tall among his peers as a respected taxonomist, a man committed to solid research, discovery and precise science. And, more broadly, as a detailed looker into the botanical matters which affect us all. A person with a minute understanding of the botanical framework which controls all life on earth.

And this too is Kew. There are over 120 scientists on staff busy researching and advising on anything from expiring species in the rainforest (many re-propagated at Kew) to drug testing for HM Customs, pollen dating of priceless antiques, medical data on plant poisons or suggesting a good crop for desert cultivation. This is the secret garden, less well known and understood, but every bit as important as the gaily coloured plant beds or luscious hothouses.

Iain is an uncanny mix of statesman and scientist, manager and missionary, visionary dreamer and practical researcher, jungle explorer and cloistered academic.

He also has Anne, his wife. Without her, he freely admits, he would have been very few, if any, of these things. And if any should doubt the reality of this, let them stop anyone at Kew and ask who they feel draws their sense of the place together. I guarantee Anne will be mentioned in the same breath as Iain, every time.

The royal gardens at Kew were taken over in 1772 by King George III on the death of his mother, Augusta. They were based on her own small botanic plot on which she had already built up a considerable collection of plants. George III may have lost Britain the Americas but certainly deserves recognition for this act alone. For this patch of London (far out in the fields of royal Richmond in the king's time) is now a tranquil pool of green amid the concrete and plastic of the modern city.

Of course London has its parks, but Kew is special. Not only because of its unexpected tranquility (aircraft aside), and horticultural authority in a nation known to be fanatical about gardens, but because of its remarkable history and contribution to an old imperial splendour.

In 1840 Queen Victoria gave her royal gardens at Kew to the nation and so the Royal Botanic Gardens, Kew, were officially established and William Hooker appointed as its Director.

Kew saw the flowering of Victoriana, literally, and alongside it the establishment of new, elegant natural sciences, practised with polished brass instruments over tea and cucumber sandwiches. Kew is the mother of all botanic gardens and it still retains this sense of period, of knowledge being slowly and painstakingly acquired down the years. Of hours spent investigating plants, discovering, naming, gently understanding them, teasing out their secrets. Today the instruments are no longer brass, and the practitioners wear white coats not knee breeches, but there is still a great sense of 'plant power' churning away in these quiet royal gardens.

As an agency for British colonial expansion Kew has had its fair share of botanical intrigues and high adventure. In the nineteenth century, paying threepence for every sample of flora collected worldwide, Kew underwrote the livelihood of many a distant Victorian adventurer – not least the Amazonian botany pioneer Richard Spruce in whose footsteps Iain himself was to follow a century or more later. This was all to the good for botanical science and the commerce of the Empire.

India, South America, Asia – from everywhere the imperial fleet touched, discoveries were sent to the growing depository of plants at Kew, dead or alive. The 'herbarium' for the dried specimens; the garden beds and glasshouses, including the famous and magnificent Palm House, for the living. From this early legacy has grown the finest collection of flora in the world, with over 40,000 different plants still cultivated on site.

Today the herbarium register records visiting researchers from dozens of different countries every week, all coming to look at some part of the incredible *six million* preserved plants.

Captain Bligh's *HMS Bounty* was carrying breadfruit for Kew in 1787 when the commander showed that he rated his cargo rather more highly than his crew by giving the plants too much of the precious fresh water aboard. The most notorious mutiny of all could be said to be down to Kew. Kew gardeners (sailing with an older and perhaps wiser Captain Bligh, who survived the mutiny) later did oversee the first planting of breadfruit in Jamaica and the Caribbean. By the middle of Victoria's reign Kew had sent over 5,000 seed packs and 2,500 living plants to the Empire: to India, Ceylon, South Africa, New Zealand, Australia, Mauritius, even the lonely Falkland Islands. Kew broke the Brazilian hold on rubber by sending plants to Malaya where it could grow on plantations free from Amazonian diseases. Kew advised on coffee plantations in Kenya, and helped establish the (quinine-producing) Cinchona in the Far East which brought relief to millions suffering from malaria.

Through its imperial mandate, Kew drove the understanding of botany forward under the patronage of figures such as Sir Joseph Banks, horticultural advisor to King George III, who sailed with Captain Cook in the *Endeavour*, and later Sir William and Sir Joseph Hooker, names writ large in the history of natural science, the first two official directors of Kew.

So, over the years, with active royal help – the estate was nearly doubled under Queen Victoria – Kew has done much in the name of science, has striven to educate and enlighten, administer and advise, co-ordinate and record all that pertains to plants. Curious, complex, and to some seeming a little out of time with the imperial age now over, Kew is still urgently active, undertaking vital work which will surely be needed in the next century even more than it was in the nineteenth and twentieth.

The present director intends to make sure that it is. And, as I worked alongside Iain on many occasions following our first meeting, the feeling grew and grew that the gardens and the man were not only busy about the same business but also, beyond that, somehow had a strange concurrence, almost like two characters with the same background, the same complex personality.

Hearing Iain lecture, for instance, is a remarkable experience. For a start he is no flamboyant raconteur. On a platform speaking with slides the man is master of his subject, offering information to the audience in a series of fact bites. But he is not a skilled public speaker in the accepted sense. He delivers largely in a monotone, moves at breakneck speed from point to point as his chosen slides dictate (he never gives the same lecture twice) and scarcely raises his voice though clearly passionate about the plants, and the urgent need to be concerned about their conservation. Yet he is in constant demand across the country and around the world. He refuses four out of every five requests to speak.

When on the staff of the New York Botanical Garden, where he worked for twenty-four years, he was once called upon at short notice to talk for 'just half an hour' on economic botany to a group of women from the Manhattan South Horticultural Society. After his talk there followed a whole hour of questions and answers. Then, over subsequent days, dozens of letters began to arrive at his office offering other speaking engagements or following up points made. For weeks afterwards his American secretary, Mickey Maroncelli, fielded calls and declined offers. And, she says, this was not at all unusual.

Even today speaking privately in closed conference with visiting heads of state, international business magnates, ministers and princes, his cool, distinct manner and complete command of subject weaves a magnetic spell over them all.

This is so like Kew itself. For Kew is not a show-piece garden. There is just too much packed in there for lavish style, fashionable floral artistry or magnificent landscaping. It is rather a place which demands active, intelligent interest. It is still a garden of great beauty with superb set pieces and unique architecture, but when you visit Kew, there is nothing flashy, nothing superficially entertaining to grab you. If you love plants, the more you look and inquire, the more there is to find. And then its subtle richness becomes slowly, almost awesomely, apparent. And further, for the gardens and for Iain, behind the commanding public face lies the vital science.

If you were to sneak down to Kew Green at about six thirty on a weekday morning you might just glimpse a lone figure slip quietly out of the director's house, a porticoed Georgian building originally built for members of the royal household. The figure moves briskly through the soft pools of light cast by the wrought iron street lamps, crosses the small triangle of grass in front of Kew Main Gate and disappears into a modest building discretely marked 'Herbarium'.

This is Iain's private time as a scientist, before his business as director crowds in upon the day. Here, the solitary enthusiast becomes engrossed in his laboratory, making notes, waggling his beard in delight, muttering occasionally in Latin and fussing over pile after pile of dried brown plants pressed out in newspapers.

He is checking, classifying and re-classifying the collected pressed specimens that are sent to him from all over the world for identification. Plant families, such as the Brazil nut, on which he is a world authority. Sometimes he tut-tuts about an incorrectly applied field label or sloppy reference, or smiles in triumph as a familiar feature shows up and a piece of the complex key to an unknown plant's identity fits into place.

This is basic taxonomy, the accurate naming of all plants, their families and species. Even today the best taxonomy is still done by close observation, with the help of lens and microscope, just as the Victorians did it. It is a lone, fascinating science and one of the few still accessible to the skilled amateur, though rigorous in its professional compass. Now, as with so much today, taxonomy is assisted by computer analysis (pioneered by Iain himself in the sixties) and identification is confirmed by DNA scanning, bark and root dissection, chromosome analysis and other modern methods. But the essentials are still the same and the boyish professor, with his stoop (real this time) and pursed lips, loves it all. Musing over the fragile flower sections in the microscope or pausing to lift the whole dried plants, brown and wizened like tobacco leaves, in and out of their paper folders, he acts with the loving care of an antiquary handling a priceless tome in a Bloomsbury bookshop.

The world outside is forgotten. The master is at work, and the work is exacting. You know this instinctively, but if there were any doubt, a glance up at the shelf full of 'monographs', each one five-hundred close printed pages of detailed reference material on an individual plant family, dispels this. They are kept there in case details need checking, though they rarely do. He wrote them all and the information is already in his head.

The Kew Herbarium is the envy of the botanical world, a comprehensive library of dried leaves, flowers, bark, fruit and roots from all over the globe. Six million specimens, collected and dried by thousands of field botanists since the eighteenth century.

And who sends him so many dried plants? The leaders of field expeditions, sent out from Kew and from botanical centres the world over. Iain knows a good deal about expeditions too. He has led dozens himself over the last thirty years. He has flown, driven, walked, paddled, climbed and, above all, collected plants over thousands of miles of primary Amazonian rainforest, not to mention mountainside and savannah. He has been stranded by plane crash, travelled for days through the jungle in a high fever, lived with various Amerindian tribes and spent hours at a time, at night, standing chest deep in snake ridden-swamps, flashing his torch in the eyes of passing caiman alligators, monitoring the pollination of the giant water lily, *Victoria amazonica*. 'How else,' he asks simply, 'could I be quite *sure* of my facts?'

A fitting piece of research for a future director of Kew Gardens. For *Victoria amazonica* has held pride of place in the hothouse for over one hundred years.

2. CHILD OF THE ISLES

Ghillean Tolmie Prance was born in the village of Brandeston, Suffolk, in eastern England on 13 July 1937. His mother, a Scot, intended to give her son a permanent Celtic perspective on life, and bequeathed him her own distinguished family name – Tolmie – and the unusual forename, Ghillean. The Scots accept the name as normal but the rest of the world tends to misread it as the English girl's name Gillian. The young Ghillean soon tired of receiving letters addressed 'for the attention of Miss G. Prance' and, with the family's tacit approval, adopted the equally Scottish but less confusing 'Iain' for general use.

The Tolmies, Iain's mother always said, were shield bearers to the chief of the clan MacL. No one was happier than she when, three years after Iain's birth, the family moved north, following a brief sojourn in Somerset, to the Isle of Skye to see out the war.

It was Iain's father, not his mother, however, who was instrumental in the move which ensured that Iain early became acquainted with 'real' nature, as distinct from the tamed and packaged kind available in towns and cities.

To most who met him, Basil Camden Prance was a grave and solemn man. But those who knew him well discovered the shy sense of fun which showed through only when he relaxed with his family and close friends. Neighbours in Brandeston were mildly shocked one Christmas to find him unexpectedly and mischievously plunging into a game of charades – as a mole, digging himself energetically under the sofa. It seemed to them quite out of character in this tall, seemingly aloof man.

Basil Prance had been for many years an official of the British colonial service, working as a magistrate in India. Propounding the law to innumerable miscreants over the years had given him a sombre demeanour. The law was the very fabric of Empire – no matter for frivolity. 'Judge Prance', as he was universally known in Suffolk, retired early from his civil duties, as was usual in colonial service, and bought a farm in England. Retirement and the return home also gave him the opportunity to consider marriage.

He had met vivacious Margaret Macnair in India. He first knew her as the wife of a lawyer friend and colleague. But his sudden death left Margaret a widow at the age of twenty-nine, and she had a son, Alick. Formidable Judge Prance, the retiring and crusty old bachelor, had fallen in love with this brave, outgoing Scottish widow who refused to let circumstances daunt her spirit. She in turn responded warmly to his steady, mannered charm. Married in the summer of 1935 they remained a devoted couple.

Judge Prance had long determined to return home to England to take up a second career as a farmer. But, as he settled into rural life – missing perhaps the colonial grandeur he had once known – he found he lacked a long-term interest in farming. So he moved with his wife and Alick and now young Iain to a vast and impressive mansion across the country in Somerset. Here her third child, Christine, was born.

By this time the Second World War had begun. The new house quickly proved too large for practical living. Too old to enlist for active service, but determined to serve, Basil Prance learned, through family contacts, that the local defence volunteers, the famous 'Home Guard', was in need of organization in the Scottish island of Skye. The family moved north forthwith.

So Iain's first remembered home is Dunvegan House, overlooking Dunvegan Loch on the north-west tip of Skye. Dunvegan House was effectively the 'second' house in the tiny village. The first, the Laird's residence, was Dunvegan Castle, the oldest inhabited castle in the British Isles, seat of the MacLeods. But Dunvegan House, unlike the castle, had no battles to withstand except the relentless battering of the elements. The solid grey building faced the sea loch, whose fishing boats and grey seals were often eagerly watched by toddler Iain from his bedroom window.

It was a particularly secure and secluded existence for a boy in wartime Britain, far from the bombing of cities in the south and insulated from the privations of food rationing. Skye was a long way from anywhere, and the locals had long learned to be self-supporting, living on what they could grow for themselves. So rabbit from the moor and mackerel from the loch supplemented vegetables from the large kitchen garden adjacent to Dunvegan House. A family cow, kept on the croft, provided milk and butter. With his local defence volunteers Basil kept a wary eye on the countryside (and on the kitchen garden which he loved to tend), whilst following the distant war through broadcasts received on the imposing valve radio in the corner of the lounge.

Iain was regularly taken out onto the loch by his parents. His first fishing expedition caused him some alarm, and them some merriment. From watching his father effortlessly draw in his line when a fish had bitten, Iain had concluded that fishing was a pretty simple business. But when his own small float began to duck and weave with his first fish, the pull on his rod was enough to make him scream out loud in terror, until his father leaned quickly over to help him secure the prize. Iain had learned an early lesson that in return for enjoyment nature demanded respect.

His mother also took to the loch alone with her little son and, her spirit lifting with the grandeur of the scene as she rowed, she would break into song, in Gaelic. Songs were sung at home too, as she sat by the roaring fire, lit for much of the year against the chill of the climate. Ancient songs from work and play: milking, churning, caroling, herding. The children – four now, since little Adine was born soon after the move – came to know them well. But out on the loch there was one special one she would always sing: a song for the seals. And as the sleek grey heads darted in and out of the water, around and about the drifting boat, she would sing out the lonely lilting song of the sea coast that was the Celtic call to these boisterous creatures. It was a magical childhood memory that Iain would never forget.

The Tolmies had in fact had much to do with preserving Gaelic songs. Frances Tolmie, Margaret's great-aunt, had supplied the avid folk tune collector Marjorey Kennedy Fraser with most of the Gaelic folk music she acquired. Frances herself had diligently collected and noted it before the habit of passing these songs on to the next generation died out in the highlands and islands.

It is easy to see how young Iain began to develop a love of nature on Skye. The power of it, compared with the puny efforts of human beings, was evident in what he could see, even from the security of his bedroom. It was not only the thrashing terror of a caught fish that impressed itself on his mind at that early age, but the fury of Atlantic gales, blowing in unhindered from the shores of America, rattling the window panes and throwing up jets of spray against the small wooden pier down on the shore where the boats lay.

In 1943 the ocean was a battlefield on which the survival of Britain depended, and those gales were a welcome relief to the sailors struggling across it. But to Iain, despite his father's interest in the ebb and flow of war, and the inevitable family analysis which followed all news bulletins, it was the fury of the wind and waves and the might of them, beating back against the shores of the loch, which drew him. That and the chance to wander at will out among the heather or down by the shore, safe from the dangers of city life, in peace or war. His own battles, when they came, would be fought to preserve the natural world he first saw through childish eyes around Dunvegan Loch. That same loch, peaceful and calm on a summer's day, was also a feast of nature: fish, seals and sea-birds, sea-urchins and seaweed, and coastal rock pools of the sort that all little boys like to play in. But soon the time came for some of the play to stop and formal learning to begin. For Iain it started, as so much informal learning did later, at the knees of older women.

Indeed his whole upbringing was shaped by the concerns and cares of a regiment of ever-present females, from his talented and energetic mother to countless cousins and aunts (and loving sisters) whose influence on Iain was as considerable as that on Sir Joseph Porter in *HMS Pinafore*.

The first of these middle-aged women of unstinting kindness were the sisters Meg and Zella, cousins of his mother. Official 'maiden aunts', who lived at the other end of Dunvegan village, they taught Iain and a small friend, the son of the only hotelier in the north of the island, the 'three Rs', every weekday morning.

They were good at their job, and Iain remembers his time there as enjoyable. The daily walk from Dunvegan down the long road into the village, past the half-dozen neat houses and up two steps into the sisters' parlour was the cosy start to a disciplined learning process which over the years Iain was to blend so successfully with the enthusiastic learning he did naturally and earnestly alone. Good science needs a well-trained mind, as well as enthusiasm. Both were begun in Iain on the island of Skye.

However, their home in Skye lasted only until the signs were clear that the war would soon be over. Just before VE Day the family said grateful thanks to cousin Anna McKenzie (Iain's godmother) at Dunvegan for her hospitality, and moved south again – to England and the village of Toddington in Gloucestershire. It was a typical village in rural England, with manor, church, school and public house. It would be hard to find a better representative of the idyll which so many English people dream of when abroad, and this may have been just what persuaded Basil to choose it as home.

Shetcombe House, where they settled, is a modest red-brick detached building set on a hillside among fifty acres of field and woodland. In those days it overlooked vast orchards of plum and apple trees across the vale of Evesham, the fruit basket of England. It is away from the village proper – a dignified, appropriate distance away for a family of standing, but not so far as to be aloof. The Prance family was a part of the community. In fact very soon it seemed that Margaret Prance was running a good deal of it.

She was certainly someone who liked to take part in things and her name was soon added to the lists of parochial church council, school governors, the women's institute and the pantomime committee to name but a few – this last in the role of producer and playwright, for she had already had poems and plays published under her first married name, Macnair. Since she also ran the Sunday school (a purely temporary arrangement, as she explained, though she was still in charge twenty years later) there was a ready-made cast list for the chorus; and being of a practical nature, she rewrote the pantomime each year to accommodate the varying number of available adults.

This was fifties England. Iain wore flannel shorts and went to church twice on Sundays. Steam trains whistled shrilly in the distance on branch lines running sleepily to Cheltenham. There were baths and Enid Blyton before bed. Once again Iain's bedroom presented him with a panoramic view – this time across the serried rows of fruit trees to the Malvern Hills, outlined on the horizon twenty miles away. Closer, framing the scene, the lower slopes of the Cotswolds stretched out either side of the hill on which the house stood, pointing westward in perspective lines to Malvern College, the school he was to grow up in and, much further west, the New World, where he would carve out his career.

And then there was the woodland, almost fifty acres of it, hard by the house, with endless opportunities for camps and dens and ambushes and hunts and adventures. For children without television this was heaven.

The house and gardens provided space to play in when it was wet and acres to roam when the sun shone. Shetcombe became a magnet for most of the children nearby, for there was always a warm welcome from Margaret, and lively friends, Iain and his sisters (Alick was by now in his twenties), to play with.

There were, however, dark clouds on the horizon. Basil was not in the best of health and, as their second year in Shetcombe drew to a close, lung cancer was diagnosed. He had always been a heavy smoker. Margaret broke the news gently to the children. She kept nothing back that they were able to bear. The certain knowledge that she would shortly be widowed for the second time was doubly hard for Margaret herself.

Iain was sent off to weekly boarding school in Cheltenham at the age of eight. He hated it. In the sorrow which surrounded his father's decline he was mainly left to fend for himself. The school itself was largely to blame for the poor results which added to the sorrow of the young child as the family did what it could to prepare for the approaching tragedy. Only the old family friend and nanny 'Woody', who lived nearby and came in to help Margaret with the children, seemed to notice Iain's mute appeal for help at the weekends when he came home. Woody (Mrs E.B. Wood) knew children well, having cared for other people's – all members of the Prance family – all her life. Whatever happened, Iain's future must not be sacrificed to the present distress. Seeing the problems lay not in the boy but in the circumstances, Woody took action.

She drew the attention of the distracted Margaret to the change in Iain and Margaret, in turn recognizing the truth of the old lady's comments, distanced herself sufficiently from thoughts of her now bedridden husband and investigated. As a result, Iain was moved to Seaford Court, a preparatory school in Malvern. Iain's work improved immediately. Woody had been right.

Within a short time his father died. That the family came through it in the way they did was due to the remarkable courage and resilience of Margaret Prance. Widowed twice, first at twenty-nine and now at forty-eight, she rallied once more. Drawing close her children, and an increasing army of visiting aunts, she set about taking on life once again as a single mother.

3. A GROWING PASSION

The woodland by Shetcombe House, split in two by a large field, became for a time a refuge for Iain. To him as a young boy his father's death was incomprehensible, and it hurt him deeply. To lessen the pain he threw himself more and more into an interest which was rapidly becoming a consuming hobby: natural history.

The woodland adventures continued. But in between charging out on the enemy – as often as not sisters Christine and Adine – he would stop, look about and begin to wonder at the wealth of interest in the woodland world around him: the neat, nested structure of an acorn, the narrow dog-like tracks of a fox, the piping, insistent alarm of a song thrush, the tenacious grasp of ivy onto a tree trunk. All these began to challenge him with questions and fire a passion to know more.

His investigations were not at first a quest for direct answers – it was more that he felt an increasing seduction by this suddenly complex world, its inter-relations and clever mechanisms. It was a fascination which grew in much the same way as some boys want to know how a car or an aeroplane works. Iain took an interest in these too, but in the woods and fields around him found a richer resource for his hungry mind. And the more his mind was occupied, the easier it was to bear the loss of his father.

He did answer many of his questions. He learned to watch the natural world go about its business while hidden by his rough 'hide' of laurels. He read reference books on birds and plants. He made notes on what he saw. And he collected. His 'nature museum' in an old box-room on the top floor of Shetcombe House was something to behold. Here he laid out the fruit of his labours. Permission had to be sought – and expressly given – before anyone but himself was allowed to show visitors around it. It was not so much that other hands might damage the careful displays, but that the wrong information might be given out about the items on show. Of course, anyone who came to the house with Iain had no choice but to be taken up and shown around.

Here, on rainy days, Iain would sit, cleaning, sorting, labelling, noting and arranging: pressed flowers, birds' nests, insect larvae, leaves, twigs, rabbit skulls, fossils, pine cones, stones and old bones. It was his train set and balsa-wood aircraft model in one. Slowly, gradually, the pain of loss eased, and the sun began to shine again on Shetcombe House.

A book which particularly captured his imagination at this time was Gilbert White's *The Natural History of Selborne*, a record of natural observation written by the parish priest of a small English village in the closing years of the eighteenth century. The book remains a classic of observation and of practical science performed without the use of complex instruments, and compiled without the author needing to move outside the parish boundary. Iain was enthralled by the detailed, accurate observation in it and immediately set out to look for hedgehog droppings!

Gilbert White had found, surprisingly, that hedgehogs ate beetles – an unexpected diet. He knew this not by employing concealed cameras or using radio isotopic tagging (as might be done today), but by close inspection of their droppings which revealed crushed insect shells. Clear scientific evidence. Iain was delighted with this straightforward piece of science and had soon tracked down some droppings himself. He checked and White's analysis proved to be correct. His mother, though, drew the line at making the evidence a permanent exhibit in the Shetcombe attic!

Also supporting the cause of natural history were the 'Prance aunts'. A formidable and slightly eccentric, but wholly welcome, band of older women who came and went in Shetcombe at Margaret's invitation. There were four: Jessie was a nun, very short at 4 feet 11 inches and sweet; Hilda was from Oakhampton on the moors of Devon, and the rather deaf widow of a naval captain; Dora, from Essex, was a manic gardener; Gertrude, a former missionary in India, had no home in England and so always arrived with her own bed roll.

Their precise relationship to the family was not quite clear, but to the children they were 'aunts', and brought colour, personality and not a little embarrassment into their lives. And all of them knew a great deal about plants. Much of this they had learned from Iain's great aunts, Charlotte and Alice. Charlotte had been a noted flower painter and Alice had made a magnficent collection of grasses, which eventually came to Iain.

Dora and Gertrude were the most frequent visitors. They had the unusual habit of keeping their money in a sort of purse slung from their waist underneath their skirts. It was no doubt safe from loss, but in church on Sunday the family (and congregation) had much trouble biting back their laughter when the aunts began to wrestle, rather immodestly, with their attire in order to find some change for the collection. And Hilda, who had problems hearing most things, would boom out, '*What* did he say?' intermittently throughout the sermon. Her strident tones had been cultivated on Dartmoor, where, since the horn on her car didn't work and sheep were a constant hazard, she would lean out of the car window and shout 'Oink! Oink!' in a forceful manner to clear the road of livestock.

For all their eccentricity, the Prance aunts' love for the children and their genuine expertise as amateur botanists were just what Iain needed. They could answer the questions his own research could not and point out ways to progress his enthusiasm. 'Going out botanizing' became a favourite pastime – though Iain's sisters sometimes needed a little pushing to embark on these expeditions. Christine particularly, usually preferred to read quietly at home.

There was a certain wild fun, not to mention bizarre fascination in going out botanizing with an aunt. The unpredictability of it was exciting. So, when called, all the children usually went. Adine, the youngest, became quite good at remembering the Latin names of the various plants and would recite them out loud. They sounded great fun to her: '*Epipactis purpurata, Veronica beccabunga, Tragopogon pratensis!*' she would trill out loudly, as though producing the names of a troop of dwarfs from a familiar fairy story.

Iain would help his little sisters over stiles and across streams whilst the available aunt (they were all in their seventies) would plough energetically on across the Gloucestershire countryside, extolling the magnificent displays nature had to offer.

Behind it all was the support and encouragement of the children's mother who often helped extend these expeditions further afield by driving them all out in the family Ford 8 to selected distant locations to spot birds, watch animals or examine plants.

Woody, who watched over the children as informal nanny, also lent support and showed pleasure at all their enthusiasms, large and small. Increasingly she stood in as housekeeper when Margaret was called away on other family business.

Warm and dependable, Woody was loved by the children, though she had one failing: she possessed only modest skills as a cook and would fasten rather too readily on any child's stated favourite as a sure road to culinary success. Once, at the start of an Easter holiday, Iain mentioned that he 'absolutely loved' chocolate blancmange (a delicacy unheard of at boarding school). Sure enough, Woody dutifully churned out chocolate blancmange every day for a month, even after Iain had left and gone back to school! The only one to complain (very quietly) was Christine, who could never stand milk puddings of any sort.

But Iain was not brought up entirely among women. There were the masters and boys at his school, and at home in Shetcombe a local member of the landed gentry began to take a hand in his education. The Honourable Guy Charteris, the younger son of the Earl of Wemys, was a great amateur ornithologist and botanist. He was also an occasional actor, and this was his primary connection with the Prances since he took part in many of the plays and pantomimes regularly produced by Margaret for the entertainment of the village, and, at Christmas and Easter, for the edification of the church.

Guy's manner of dressing off-stage generally intimated this blend of gentle aristocrat, field expert and artist. A black beret was always on his head, and half-glasses, which he peered over in a studious but friendly manner, underlined his eyes. A tweed jacket, trousers and heavy walking boots completed the outfit. His close acquaintance with nature was what Iain most admired. Guy's knowledge of the bird life of Gloucestershire was unparalleled and few holidays would pass without them doing some bird-watching together. He also often took Iain to see his old friend Peter Scott at Slimbridge Trust wildfowl sanctuary on the Severn estuary. There, under their joint tutelage, Iain wandered freely around both public and protected areas watching and learning about the bird-life and gaining an early lesson in conservation.

The time came for Iain to move on to secondary school. By the age of thirteen he had already been at boarding school for five years, returning home to Shetcombe for holidays and the occasional weekend. And since private boarding school education was considered the only appropriate kind for him, at thirteen he was sent on to board at Malvern College.

Malvern College was, and still is, an exclusive private school. It is set in spacious grounds at the foot of the Malvern Hills, and in Iain's day was home to some six hundred boys (there are now girls as well), aged thirteen to eighteen. Close behind are the Malvern peaks: Worcestershire Beacon (1,394ft) and Herefordshire Beacon (1,114ft). The school itself consists of a dozen or more gracious Victorian buildings, including a private chapel and refectory, individual boarding houses, classrooms, science blocks and language centres, and a music school housed in an old monastery. To the east of the buildings are extensive cricket and football pitches and to the north is the spa town of Great Malvern itself. It is in every way a rich environment in which to learn.

In summer, strawberries are eaten on the headmaster's lawn to the clip of leather ball against willow bat as the first eleven pursue their leisurely but earnest game against an opposing school team. In winter gaggles of children stamp their feet on frosty paths, exhale smoky clouds of vapour and wind their bright school scarves tightly around their necks before rambling over to college chapel for the traditional Christmas Nine Lessons and Carols service. Malvern College is British establishment to the core. It was to be Iain's home, for nine months of the year, for the next five years.

Margaret inspected the school herself, with Iain, before deciding on the move. There she introduced him to the man who was to be his housemaster and who, in the event, was to become more significant than any other in guiding him to his eventual choice of career.

Bill 'Basher' Wilson, housemaster of No.1 House, overseer of around a hundred boys, boxing coach and master of biology was tall, fit and commanding, with steel grey hair and a brisk manner. His nickname derived from his sporting duties, but he was gentler than the name suggests. 'Botany Bill' would have been more appropriate, for although his academic discipline was biology, his real passion was field botany, the close study of plants and their habitats. He was an expert on the flora of Britain.

Straight away, on this first visit, Bill Wilson noticed the boy's interest in the school museum. This fine collection of artefacts made Iain instantly aware of the paucity of his own collection back at Shetcombe.

As his mother talked of term dates, tuck boxes, sports and studies, Iain patrolled up and down the avenues of glass cases, spotting and commenting, wholly fascinated. Bill warmed to him immediately, remarking to Margaret on the boy's inquiring mind.

Iain for his part already knew for a certainty that if his mother agreed to send him he would have a great time at a school which kept a museum like this. And although to start with he endured rather than enjoyed the crude and rough initiation of a 'new bod' at public school, he settled in quickly enough. His enthusiasm and ready interest in people and ideas, his quiet resolve and his complete disregard of what others might think of him soon won the respect of his masters and peers. In due course he rose to be head boy of No.1 House, although he never particularly shone at school, being disadvantaged at sports because he wore glasses, and finding the rewards of study only through repeated effort.

Iain took his Latin O level exams no fewer than three times. But because he was determined to get to one of the Oxford colleges (his mother expected it, it was a family tradition) and Oxford required at least Latin O level he kept on until he passed. In doing so he learned that persistence, not just brilliance, can often earn a coveted prize – something he had cause to remember many times in the course of his career.

Malvern was where he made that career choice. Not overnight, for it took the full five years to make. In the end his latent interest in nature, informally coached at home, developed into such a consuming passion at school that the choice was obvious. But the fact that this passion had focused much more on plants was mainly due to the personal interest of Bill Wilson, his housemaster.

Bill was a bachelor and had developed the habit of setting off on expeditions at weekends and in school holidays to indulge his own botanical enthusiasm. On these trips he would often take along boys who showed a similar interest. Very soon Iain was one of this privileged group and showing such an ability that Bill recognized it as remarkable.

'He was just so single-minded, not narrow, just single-minded,' he recalls. While the others, no less keen, would enjoy the walk, ask questions or notice the plants, animals and birds, Iain would be plunging ahead through the undergrowth to investigate tree, flower, fruit and leaf with exhaustive precision. He made the subject, and often the expedition, his own.

Very soon their required companion on trips was a weighty tome entitled *The Flora of the British Isles*, edited by Messrs Clapham, Tutin and Warburg, a comprehensive guide to the plants, common and rare, likely to grow in the British Isles. This was occasionally supplemented by others including *The Flora of Gloucestershire*, a book of Iain's, given to him, inevitably, by an aunt. Both proved comprehensive guides to the green treasure to be mined in and around Malvern, the Severn Valley and the Vale of Evesham, the group's main botanizing country.

Iain soon learned not only the names of local plants (some of which he knew well already) but also, step by step, how to find his way through the 'keys' to family, genus and species of plants he did not know. 'No pictures!' commented Bill Wilson firmly. 'Keys are precise descriptive terms.' So Iain found. They were coded one-liners such as, 'Leaves opposite, ovules attached to the wall of the ovary' or 'flowers with staminodes between the fertile stamens'. Once mastered, these crisp designations narrowed down a chosen specimen to a few options among plant families (108 in the British Isles) and then on down through further sub-classifications to confirm genus and species, of which there are about 1,500 in the British Isles.

This reductive process, demanding a minute examination of flower, leaf, fruit, stem, texture, overall shape and plant colouring led, in the end, to a final choice between just one or two individuals. Habitat and location provide clues as well, and so a firm identification is made. At this stage a grunt of confirmation usually came from Bill, accompanied by a click from his Retinet portrait camera, and the little expedition would move on to start the whole process again at another plant site.

Gradually familiarity with both plants and keys meant that the answers came more quickly and Iain built up expertise with the basic tools of what was to become his chosen profession.

As the terms rolled by and the holidays came round, local botanizing with aunts gave way to trips further afield with Bill in his car. There were particular quests for rare or unusual plants. Word would reach Bill or Iain that a sought-after rarity was to be found in a certain railway cutting near a station in Wiltshire, or atop a particular escarpment outside Hereford. Early in the morning, Bill's old grey Vanguard would wheeze up the Shetcombe hill and Iain would tumble out to greet his housemaster and a school colleague or two, armed with sandwiches and thermos, ready to scour the target area. Plant information of the kind they acquired was, and is, often kept very secret, with only a photograph proving a successful find for fear that too much attention will eliminate a barely lingering species. But successful finds were nonetheless often made by the Malvern College boys.

Adonis annua (pheasant's eye) was one such elusive rarity, all the more difficult to find since its leaves so much resemble the ordinary buttercup. The search for it occupied a number of weekends and the trusty Vanguard ranged further and further afield looking for the appropriate 'dirty cornfield' (poorly weeded) habitat. Adine came along too, though increasingly in her teens she was perhaps less concerned to look outside the car for her idea of Adonis than inside, at Iain's school friends. In the end they found the plant by a newly-built military road crossing Salisbury Plain. The ground broken by the sappers had enabled dormant seed to germinate in the topsoil. This was another significant lesson for Iain for the future: new roads frequently offer the botanist a chance for rich pickings.

But expeditions aside, one of the abiding memories of the Shetcombe days for all the children was also of flowers – spring daffodils. Golden hosts of these massed each year in a sloping five-acre field opposite Toddington Manor, on Prance land, which divided the woods in two. They had originally been planted by a former owner of the manor, an invalid who had wanted a display she could see without leaving her bedroom. The idea had outlived its owner and in the Prance's time the daffodils blazed out not simply for the manor but for the whole village, which was just how Margaret thought it should be anyway. So she charged villagers a shilling a time (sixpence for children) to pick as many as they wanted for their homes and cottages, and used the proceeds to buy new pew seats for the church or to support cancer research.

The true flavour for Iain and his sisters of their childhood is summed up in that time: the wealth and beauty of nature, the care and community spirit of their mother and spring joy returning to the family after bereavement, and the ever-friendly presence of the solid red-brick house on the hill looking out across the valley to Malvern.

4. DREAMING SPIRES

Much as Iain enjoyed learning about plants and going botanizing in his free time, the main business of school was to get results in more conventional subjects. No one was convinced that Iain could make a career out of his hobby, nor indeed that he had the mind for it. It had been a struggle for him to get Latin in his O Level exams, and this did not seem to bode well for a life in the natural sciences. It was important that he laid a wider base for the future. So he did not even take botany at A Level: he took maths and chemistry instead since both were important as general university entrance qualifications. Oxford, which he and his mother had set their hearts on, had a different system of entry. It set its own special exams, a general paper and a specialist one.

For the specialist, he took botany and despite not having had any formal teaching in the subject at Malvern (except as a part of biology classes) he did well. He also passed the general paper. He wanted to go to Oxford and study botany. A board at Keble College met to discuss his case. An application from a student candidate to read a subject he had only studied informally at school was unusual, to put it mildly. But the exam results were impressive. Bill Wilson was contacted. What was this candidate like? Could he handle a degree course on this basis? Bill was in no doubt. A firm and positive letter went back by return: 'The boy is a natural. He will be a credit to the College.' The tutors were reassured. Iain went up to Oxford – ancient city of learning and of 'dreaming spires' – in the autumn of 1957 to Keble College.

Life as an undergraduate would, he knew, be a challenge academically. A love of plants does not translate into real scientific knowledge without much unavoidable hard work. There would be no shortcut to excellence. But his first few weeks there were to prove much more of a private and personal challenge than he ever dreamed of.

Oxford was an exciting place for a young man just released from the rigid regime of public school. Free to roam at will he could wander the narrow part-cobbled streets between high college walls of warm grey-gold Cotswold stone pierced with high arched doors leading to closed college cloisters and quadrangles, and bounded by clock towers, lecture rooms and students' quarters. Stairways everywhere bustled with undergraduates, post-graduates, dons and tutors, some in gowns preparing to take lectures, others in sports gear ready to go running, rowing or play rugby.

Squadrons of bicycles stood guard outside college gates or propped up at street corners and, as darkness fell, yellow light spilled out of the old bow-fronted bookshops and tea houses lining the streets and alleys, while inside the ideas of the day were passed to and fro in print or conversation, examined and re-examined, taken up or discarded by an army of fresh, earnest, inquiring minds.

At Oxford choices opened up in every direction, including a staggering variety of extra-curricular activities. Debating, music, drama, sport of all kinds, literary societies, art groups, science clubs, politics, poetry – the list was endless. Iain had been warned that outside activities could easily come to dominate his life if he was not careful. In fact he had already concealed from the music societies that he was a competent bassoon player, aware that otherwise his time would not be his own, with invitations to play in orchestras and bands. He did, however, continue to play the accordion, a safer instrument in this context. At Malvern he had developed an interesting stage act in which he played diverting music while a friend performed conjuring tricks. The purpose of the music was to distract attention from the magician while the crucial sleight of hand took place. Whether this end was achieved by his virtuoso performance, or sheer activity and volume, no one can now recall.

But one thing did attract his attention in these opening weeks. There was a group which studied the Bible. Iain had grown up at school and at home within a framework of Christian belief and teaching. The Church of England chapel service was a formal part of life at Malvern College, being compulsory for all boys. He had also been heavily involved in his mother's plays at home, based on biblical themes and acted out at Christmas and Easter in the church at Toddington.

Christian thinking had formed much of the warp and woof of his childhood. But as he had grown he had begun to find that much, at least for him, a matter of form and little more. In newspapers and at school there was increasing criticism of the weakness of traditional Christianity 'in the light of modern scholarship'. Iain was left with the feeling that the religion to which he had deferred, though perhaps it had held something once, was becoming less and less relevant.

Yet at this stage when so many students, faced with the excitement of undergraduate life, slip away from their traditional backgrounds or old Sunday school teachings, Iain did not. He sensed in himself a deeper need to know what was true and what was not before he rejected it all completely. At Malvern he had found a number of provocative ideas about faith in two books he had read: *Mere Christianity* by C.S. Lewis and Our Faith by Emil Brunner. And at Oxford, as he could see all around him, the business he was now involved in was the testing out of ideas. So he signed up for the Christian Union, deciding to go along to its meetings, at least for a while. This also involved going to the evening services at the University Church on a Sunday evening.

At these he was in for a shock. Instead of the straightforward, rather conventional moral homilies he was used to hearing from other pulpits, he was now subjected to a careful, penetrating analysis of the state of the world, the importance of personal conviction and the claims of Christ on individual lives. He had certainly not thought about it that way before. Here was a new vein of thought, a challenging and important discussion about the purpose of life itself.

It seemed suddenly that much of his previous thinking on matters of faith had been shallow and without substance. He had barely skimmed the surface. Within a few days he was deeply involved with the Christian Union, not so much as an active member but as an inquirer, a searcher after truth. And, as he began to digest the rich fare on offer at the Bible studies and discussion groups, he came to the conclusion that there was much more to all this than he had ever imagined.

Things came to a head one Sunday in October. The evening service at the University Church was full as usual and the speaker, Canon John Collins, spoke powerfully on the arraignment of Paul the apostle before King Agrippa (the story told in the book of Acts, chapter 26).

John Collins took as his text a comment King Agrippa made after hearing Paul speak in defence of his work and faith. In the words of the King James Bible, 'almost thou persuadest me to be a Christian'. Iain followed as John Collins outlined the arguments the apostle had used to speak to the king about the Christian faith he had adopted. It was pretty convincing. And the speaker followed it with the comment that, whatever King Agrippa had thought, the establishment of the Christian church across the Roman Empire at that time, and since then down the centuries around the world, had borne its own witness to the truth of the convictions Paul had expressed.

As he listened, Iain felt a pull, almost a call, to place his own life in the hands of this invisible but, to him, increasingly real Saviour. The authenticity of Christ's life, the claims he made, and the remarkable fact of his resurrection after his death on the cross seemed to Iain to be true, emotionally, intellectually (though there was a great deal more to discover) and personally. But Iain was a private man, a scientist, a man of considered thought. Emotion alone would never determine the way he intended to live. His decision, and he clearly saw a decision was called for, would be based on the facts. Yet this was a pull on his heart too. Suddenly he found that his heart and mind were not at odds. The whole business made sense to both. There was no conflict between reason and faith at all. It was for him a moment of truth.

The minister concluded and the final hymn was sung. Iain was in turmoil. He had not known this familiar faith was so demanding, for although church had been part of his life from childhood, he knew now that he had to make another step, a step of mature allegiance.

John Collins spoke again, inviting those who had been moved in their thinking and their hearts by God during the service to come forward and show they were able to go further than King Agrippa and say, 'I am persuaded.' Iain went. He was not alone, but it was a pretty lonely walk all the same. But, as he says now, 'It was a walk into the arms of my Saviour.'

It was the decision of a lifetime. Iain had become a Christian, and his new adult faith would determine his life from then on. Not that he became a religious zealot, far from it. But his faith deeply affected every part of his life. Now, if asked, he will speak quietly and lucidly about this faith, or discuss matters at his church, or outline his views on God's creation, or bring out a thoughtful Christian paper he has submitted at one conference or another. But only if asked. Otherwise it is only in his quiet manner and gentle patience, in the suffering of fools more gladly than perhaps he ought for one in his position, in the almost legendary calm humour he exhibits when problems arise or danger threatens, in his abiding view of nature as a careful piece of creative handiwork, not random chance, that his deep spiritual convictions show.

For the cool, level-headed, dedicated science student Iain had intended to be, that first year at Oxford was to prove a roller-coaster ride. Coming down to earth with his new faith was much helped by the Keble College Christian Fellowship which he now joined in earnest. Of particular help was a series of systematic Bible studies and discussions known as the 'New Christians Course'. This was led by the Rev. Dick Lucas, a man of clear mind and lucid exposition well able to handle the razor-sharp queries of keen-minded undergraduates.

A result of this was that Iain volunteered to help lead a Children's Special Service Mission beach mission for young people during his 1958 summer vacation in Frinton, Essex. Although it was a wrench to drag himself away from botanizing around Oxfordshire on his 1939 BSA 250cc motor cycle, as Dick Lucas told him, 'There is no better way of learning about your own faith than having to teach it to others.' So he went along. One of the other leaders on the mission was a bright impish girl with a cheeky smile and literary turn of mind. Eighteen-year-old Anne Hay, daughter of a Norwich vicar, was about to go to London to read for a degree in English. The CSSM mission was part of her summer vacation too.

Curiously, Iain found that the Bible had added impact when read out by Anne, as did a great number of other things where she was involved. It was most disconcerting. Here was yet another distraction from study which he certainly hadn't allowed for.

But Anne's company was so good. His dry humour seemed to match delightfully her more combustible sort, while her slightly mystical way of looking at things balanced his aggressive pragmatism. She could always sense the mood and see the meaning behind an event or incident, where he could usually see only the event. That was vital for work as a scientist, of course, but Anne's less rigid, more other-worldly view seemed often to complete things for him somehow.

One morning Anne learned the tragic news that her brother's new baby had just died. She had to go to Norwich urgently to be with him and his wife. Iain, who had borrowed his brother's car for the mission, was asked by the group leader if he would drive Anne to Norwich immediately. Despite the distress of the journey, Anne began to see in

Born in 1937, Ghillean
being held by his mother
at his christening

1946. With his sisters and
half-brother, Alick, who
had returned from five
years in the Indian army
during World War Two

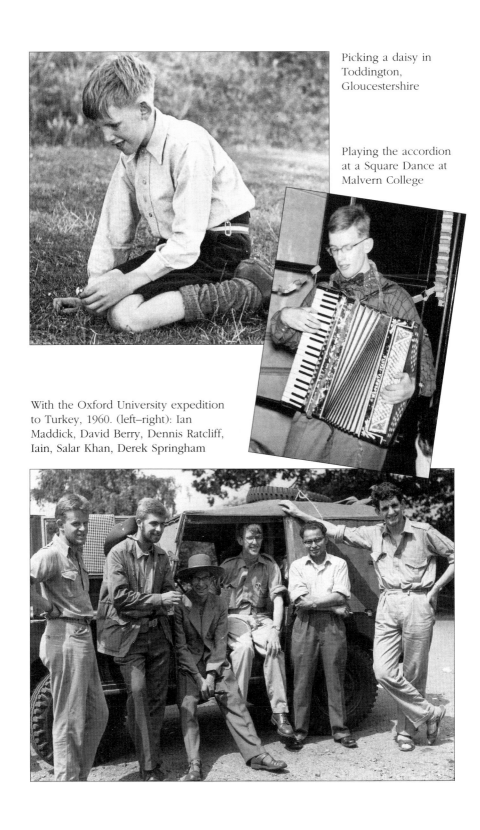

Picking a daisy in
Toddington,
Gloucestershire

Playing the accordion
at a Square Dance at
Malvern College

With the Oxford University expedition
to Turkey, 1960. (left–right): Ian
Maddick, David Berry, Dennis Ratcliff,
Iain, Salar Khan, Derek Springham

Anne Hay (later Prance) in the family daffodil field, Toddington

Anne and Iain on their wedding day, 13 July 1961

The family on a two-week excursion to Lábrea on the Rio Purus

Rachel and Sarah Prance in Ducke Forest Reserve, Manaus, Brazil

Anne and Iain in the Amazon, 1994

Iain with tree climbers in
the Amazon

Yanomami Indian
climbing tree

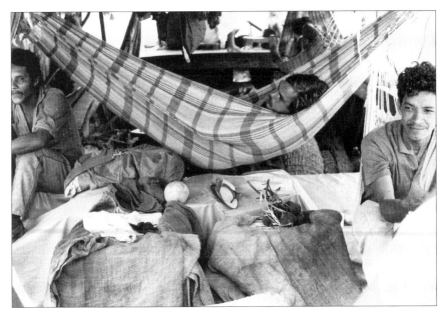

A scene inside a river launch. The launches can get very crowded on expeditions

Pressing plants on an expedition to a white sand campina forest

Samples for chemical analysis drying in the sun

A shelter for pressing and drying the samples collected during the day

A buttressed tree being cut for its timber

River rapid vegetation at Camanaus, Rio Negro, Brazil. Plants growing like this are called rheophytes

An expedition arrives among the Paumari Indians. The boat and plant specimens evoke some curiosity

Rubber boats are excellent for river-based exploration: easy to pack and carry, rock-proof and river-proof, and above all fast. William Steward is the student receiving the package of pressed plants

Robert Goodland (without hat) and students of Iain's first course in Manaus survey charcoal kilns supplying a steel factory and causing much deforestation

Brazilian students from INPA on a field trip to Mato Grosso in 1973. On the right is Eduardo Llerns, at that time Iain's teaching assistant

A rubber harvester and his family at Santo Antonio after their first Sunday service at which Iain preached

Iain in Serra Araca, Brazil, 1986

Yanomami Indian walkers setting out on the month-long walk from Surucucus to Waiká

Iain offered the Yanomami walkers a plane ride back to their village for their work; unfortunately it crashed and they had to wait for two weeks more

A Brazilian fieldworker drinking from one of the forest vines (*Doliocarpus*), which produce a delicious, fresh, potable water

Collecting on Iain's first tropical expedition to Suriname produced so many collections that it was necessary to run four plant presses simultaneously. Left to right: Frederico (camp cook), Howard Irwin (expedition leader), Noel Holmgren, Thomas Soderstrom, Iain.

The Guaraná (*Paullinia cupana*) vine with its caffeine-rich fruits is used by Indians as a stimulant and Brazilians as a cola-like soda drink

A selection of the great variety of fruit types of the Brazil nut family (Lecythidaceae) the taxonomy of which Iain studied

The monkey ladder vine (*Bauhinia* sp.)

A young Calabash fruit (*Crescentia cujete*). The surface is covered with glands which secrete a sugar, attracting ants which protect it from predators

A mature Calabash, hardened after the ants have done their work well

Iain someone of real energy and strength, of like Christian conviction and of solid purpose. She had little idea then, or even during the three years of their subsequent courtship (they did not marry until she had finished her degree at Westfield College) quite what marriage to someone *so* dedicated would mean. She could not guess then that she would set up home across the Atlantic and be left largely alone in a foreign culture, or that later she would choose to take her young family with her husband to the Amazon jungle, or that he would so frequently be away, out of touch for months at a time, while they were growing up.

'I am a confirmed anti-botanist!' she says now, with a provocative grin, when asked about her interest in his profession. But she is very pro-Iain, for their relationship has withstood more dramatic tests than most, and not only lived to tell the tale, but grown stronger because of it. Anne and Iain have complementary, not competitive, gifts which they have learned to put together in a quite remarkable way. Throughout Iain's career Anne has been there, often crucially, enabling him to progress. Initially she supported the two of them by teaching while he completed his DPhil research at Oxford (a PhD at Oxford University is called a DPhil). Later she ran his expeditions in the Amazon. Today she organizes the massive hospitality programme of the director's household at Kew (on which much of the Garden's support depends). There is a dinner or reception virtually every day.

Iain is still amazed at how good it was that he met her. And it led to one other major decision in his life. For at this time his new-found faith had led him, perhaps naturally, seriously to consider a career in the church. A number of his colleagues in the college fellowship had already decided this was their vocation, and are now distinguished church leaders. Pat Harris, Bishop of Southall and Tom Walker, a noted missionary to South America and now Archdeacon at Southall, were two of Iain's close friends at Keble.

So, despite his enjoyment of the botany course, he felt more and more that his faith should lead him to full-time work in the church. He decided to go forward for ordination. As a required preliminary for this he applied to the Church of England selection board and was accepted. Perhaps, he thought hopefully, he would end up like the Rev. Gilbert White of Selborne, an amateur botanist. But at this time of decision, during a visit to see Anne in Norwich, he began to discuss the matter with her father, the Rev. Archibald Hay, a man who knew both church and churchmen well. It was to prove a crucial meeting. 'We just talked the whole thing through,' recalls Iain, 'and he showed me that botany could be just as godly a calling as the church. He said that God would not have given me such a gift for plants if he did not want me to use it to the full.' Iain took that simple advice and was back on track, his calling in line with his heart's choice and the ability of his mind. From now on botany would be his career and his Christian duty.

He and Anne were married in Norwich on 13 July 1961, Iain's birthday, by her father. Iain was still an impoverished student, though now embarked on postgraduate work at Oxford. They moved first into a flat in Cogges Priory, near Witney, Oxfordshire and then into a small 'Bluebird' touring caravan on the farm of a friend. Anne divided her time between supply teaching, working at a joinery firm glueing headboards onto beds, and helping Iain with the computer work which was part of his DPhil thesis.

When in 1960 Iain had obtained a BA Honours, he decided to do a DPhil in Forest Botany as he now fully intended to make botany his career. This decision had been determined partly by the personalities and interests of the various lecturers at Oxford – he had enjoyed much of the time he had spent in the Forestry Department – but also by his increasing interest in taxonomy, the science of plant classification, one of the foundational sciences in botany. The herbarium of a botanical garden is the centre of such study (all dried plant collections are held there) and microscopy and simple observation lie at the heart of it. That and an encyclopedic knowledge of plant structure (plant anatomy). This knowledge is based on practical experience, supplemented by reference to specialist monographs, written by different experts in particular plant families.

It is largely a matter of detective work, noting fine detail on plant specimens gathered on field expeditions and sent to the various herbaria round the world by field botanists. When details match, all the specimens must belong to one species. But new plants are always turning up (there are 50,000 species in the Amazon region alone) and every so often a new genus is found or a family must be subdivided further to maintain a sensible and practical classification.

There is much work to be done as many, particularly tropical, plant families are only cursorily known. Superficial research has placed them only into rough categories. Even in the herbarium there is still a great deal to discover, for most have many more dried plant specimens than there are people qualified to classify them. This is often the work of the thesis researcher, or research assistant at a botanic garden – Iain's role in his DPhil days. The thesis had to be about one plant family and the work undertaken under the direction of a senior botanist. Research enables the student to learn the practical business of taxonomy on the job. A good student would also begin to provide vital insights into a little-known plant family, which he or she alone would organize and describe in detail, possibly for the first time.

Iain knew that even as a student he would be pushing back the boundaries of his science. It was very exciting. All new official descriptions of plants and families must be recorded in Latin. It is a standard requirement – botany is an international science and botanical Latin provides a common language of classification around the world. Today Iain is an expert, but as a student he found the Romans had set him a stern task in taxonomic communication!

He believes he was very fortunate in his supervisor, Dr Frank White, one of the foremost botanists on the flora of Africa. Frank was a botanist through and through, with a manner to match. His long hair and sideboards, his penetratingly exact spoken English (spiked occasionally with unexpected lapses into a native Northumbrian dialect), his ruddy complexion derived from many tankards of good Oxford beer and his complete avoidance of women marked him out as a 'don' of the old Oxford school. Brilliant and eccentric: a man to admire, to listen to, and to learn from. In addition, he spent time with his students. All supervisors were supposed to do so, but many of them did not. Iain was fortunate in getting Frank White.

Their relationship grew during the three years of Iain's doctorate studies to one of mutual respect. Frank White particularly admired the speed with which his DPhil student negotiated the streets of the university city on his BSA motorcycle. He decided to follow suit. Iain was startled one day to see the doctor bearing down on him with a rather lost and helpless look.

'Iain,' he announced in clipped tones, 'I have purchased a motor scooter, but I do *not* know what to do with it!' Hiding a smile, his student courteously led him to a nearby car park and took him through the rudiments of ridership. He later had the equal honour of teaching him to drive a car.

But Frank White was unmatched in Oxford for his knowledge of tropical plants, and from the short list offered by him Iain was quickly able to identify one family that badly needed classification. It was, in White's opinion, 'a mess'. It was the Chrysobalanaceae, a family of large shrubs and trees growing in the tropics, especially in middle Africa and equatorial South America. Iain liked the family, found that many specimens had been collected in various herbaria round the world, and made it the focus of his research.

Initially he worked in Oxford, at Kew and the Natural History Museum in London, which he visited weekly with Frank White. Then he began to call in collections from other herbaria around the world. This is standard practice in the botanical world, though it seems surprising to other sciences which tend jealously to guard their collections of materials. Kew, for example, routinely lends around 100,000 herbarium specimens around the world each year.

Soon Iain was faced with a mountain of information and material. Characteristically he chose a novel way through the maze. Both Oxford and Southampton Universities had just installed primitive computers. Could he make use of these to match up plant characteristics? It was a fairly mechanical business after all, once the specimens had been inspected. All he would be doing, in essence, was to put the collected plants into 'piles' under his chosen classifications.

He and Anne worked out thirty different characteristics for the Chrysobalanaceae of which he had the material – about 120 different identified species at the time. These were pretty standard characteristics for plants: leaves hairy or smooth, fruits hard and woody or soft, flowers solitary or in clusters and so on – exactly like the 'keys' he had learned from Bill Wilson at Malvern. These were then punched onto paper tape in binary code along with a command for the computer to match them up. The final print-out did not include a convenient graph, as it would today. All it could do, and even this was asking a lot, was to print the co-ordinates of the points for a graph. The rest was up to Iain and Anne on the caravan table.

The graphs, when drawn up from the matched data, confirmed Iain's expectations that more sub-classifications were needed to clarify the existing confusion within the Chrysobalanaceae family. He divided one large genus into six, each quite as distinct as the other existing genera. Iain felt for the first time the thrill of the scientist, the excitement of seeing his own endeavours add to the sum of human knowledge.

He also began to see the enormity of the task. So much still to discover! He looked into the bark and wood of trees, making microscopic slides of collected wood specimens, a science known as wood anatomy. He looked into the germination process – seed morphology – and identified differences between his new genera based solely on the way the first new green shoots grew out of tiny plugs in the end of their seed shell, rather than breaking open the whole. This was a fundamental difference. He looked into the pollen, into the fruit and into the environment in which the trees grew. All helped to find the correct classifications, the business of taxonomy.

He loved every bit of it. But where would it lead? Was there a real future in this after his DPhil? Frank White had a number of contacts in Africa, and a project there looked promising. But what else?

Things came to a head in the summer of 1963. Iain and Anne's first baby was born in an Oxford hospital, a strong healthy girl whom they named Rachel. In the same week Iain took and passed his DPhil 'viva', the oral examination held by the examining professors on his thesis on the Chrysobalanaceae.

By then Iain also knew that he would not be staying in the UK for long. He had been tempted by an offer which included the one thing which for him rivaled even taxonomy: a botanical field expedition in the South American jungles of Surinam. He would have to leave in two weeks. He went, for in Turkey the year before he had discovered yet another side to botany which outshone even taxonomy.

5. AN EXPLORER IS BORN

Iain had led, or rather co-led, his first expedition three years before, at the start of his DPhil time at Oxford. The official 'Oxford University Scientific Expedition to Turkey' in the late summer of 1960 had originally been conceived as a 'bit of mountain climbing'. It was right after his BA exams and he and his colleagues needed a strenuous adventure holiday at the end of their studies.

But it was at the top of a mountain on Skye, Iain's childhood island home, that the idea of turning a holiday into an expedition first came up. Iain was on a field trip run by the Botanical Society of the British Isles, of which he was an active member. Such trips were by now a common part of his experience and very much more than the old-style 'going out botanizing', though he still did this for his own interest and amusement when he could. He had determined to make use of every opportunity to go into the field, whether on Society outings or official trips connected with his course at Oxford, mainly arranged by the Forestry Department.

Iain had set his heart on specializing in taxonomy and ecology so he was increasingly involved in agriculture and forestry. He found very good tutors who were taxonomists and ecologists, and who were pleased to help and advise him. One fine botanist still in the Botany Department, though, was Dr E.F. Warburg (known universally as 'Heff'), of the famous Clapham, Tutin and Warburg triumvirate whose *The Flora of the British Isles* had formed so much a part of Iain's developing passion for plants at school. He was a keen field man and took Iain and his fellow undergraduates on botany trips all over the Chilterns and the South Downs, as well as Wytham Woods near Oxford.

For Iain, this was the way it should be done. Work in the laboratory (though for him at this stage mainly the lecture theatre) allied to work in the field. The one could not exist without the other. And increasingly the trend of modern taxonomic investigation was to identify not just the plants but the precise way the environment in which they grew fitted them. The question 'Why?' was creeping in alongside the 'What?' which had stood taxonomy in such good stead down the years.

Why did some plants have the distinctive soft floppy leaves that so distinguished them? Might it have something to do with the lack of water where they grew? Why did some have curious bell-shaped flowers? Might it be worth looking at the type of bee that did the pollination? And so on. Of course, much of this was already known as far as British flora was concerned. Many observations had already been made. But overseas it was different. Great forests of plants awaited discovery and classification, and Iain knew that many of them interacted with other plants, birds, insects and animals in unexpected and remarkable ways. Where they grew was part of the natural equation of life. His taxonomy would never be 'just plants', but plants in context, plants in nature, plants in the service of humankind, plants in the ecology of the world.

And it was increasingly evident to most natural scientists that without field work – going out and seeing it all for oneself – much time in the laboratory could be wasted. There were precious few environmental hints to be got from the dried specimens themselves. Field notes were made by all collectors, and these were vital, but on a fast-moving expedition when fifty or more species had to be collected and dried in a day (about twelve specimens of each species are usually taken), each botanist has little time for anything but the most cursory of records on the environmental factors relating to each plant. One had to be there with a taxonomist's trained eye to spot the crucial factors for oneself.

As the last year of his degree course approached, Iain had decided to try for a PhD and cast about for an institution to take him. Oxford was a possibility of course, but there was no guarantee of a place just because he had studied there as an undergraduate. There were many keen to do research at Oxford and places were limited. He had to cast his net wider. So, among others, he got in touch with Edinburgh University which had a botanical department with a high reputation. Also, if the truth were told, he quite liked the idea of the Scottish connection. Working in Edinburgh would be a bit like coming home, and his mother would be delighted.

So perhaps it was the 'skirl o' the pipes a calling' that also made him elect to go on a Botanical Society field trip to Skye that summer, even though he loved the island anyway and had been back several times with his mother and sisters since the time they left as a family at the end of the war. By coincidence, one of the other participants in this excursion was a senior taxonomy lecturer at Edinburgh, Peter Davis. Dr Davis was an experienced field botanist and an expert on European plant species, and he had a particular interest in Turkey. Unfortunately he was no longer permitted to travel there. He had been declared *persona non grata* because of a change to military government in Turkey following a *coup d'état*. The man's frustration was extreme. He was trying to compile a flora of Turkey and couldn't go anywhere near the place!

So when Iain casually mentioned his thoughts about 'a bit of climbing in the Lycian Taurus mountains' the heat began to rise among the heather. 'Young man, you *can't possibly* go to Turkey without collecting plants!' expostulated Dr Davis. 'You're *supposed* to be a *botanist* for heaven's sake!'

Iain was taken aback.

'Besides,' he said adroitly, 'people will sponsor a scientific expedition; they certainly won't a holiday!'

Iain took the point. He also quickly saw that Edinburgh University might be very glad of a researcher to work on Turkish plants if there were some to work on and especially if he had provided them.

So when he got back to Oxford he told his mountaineering companion Ian Maddick that the climbing holiday was off.

Maddick looked disgruntled. He had been rather looking forward to it. 'All that has to be organized now is a full blown scientific expedition the length and breadth of Turkey,' said Iain. 'And what's more, you and I are going to lead it!'

So Ian Maddick took responsibility for the expedition as a whole and Iain Prance was designated scientific leader. They worked together on almost everything, though Iain planned the plant collecting once they got to Turkey as Ian Maddick was a medical student with only a modest interest in plants.

Although none of the team had ever done anything like this before, the Oxford University Exploration Club had been set up specifically to steer hopeful expedition leaders in the right direction and towards the right people. It is a wide-ranging and valuable resource and is particularly good in advising on money and sponsorship. So with their help Iain and Ian began writing letters on 'Oxford Expedition to Turkey 1960' notepaper asking for support. It was the first of many times throughout Iain's botanic career. Even now, at Kew, he must still persuade people to support worthwhile botanical projects. Much botanical research is supported by grants and sponsorship funds. It generates little direct revenue of its own, though the benefits of such research may be very great. So looking for funds is a normal part of Iain's life. It was perhaps a good thing that he found out early about this side of life as a botanist.

They raised enough funds to buy two old Land Rovers. One was a long wheelbase type, diesel-engined with a hard top, ideal for the trip they had in mind, and the other a short wheelbase petrol version, less suitable and very much older. In fact they had to take the second one virtually to pieces before they set out. Iain quite enjoyed this side of things: it made a change from stripping down his BSA, at which he was highly practised. Fortunately one of their expedition members, Dennis Ratcliffe, was quite an experienced motor mechanic and they did the work partly in the garage at home in Shetcombe and partly with the help of an enthusiastic local commercial garage owner, John Goodall, an old school friend of Iain's in nearby Evesham. They got it going in the end, pretty well they all thought. But not for long.

The garage at Shetcombe housed a good bit more than Land Rovers over the months leading up to their departure. Companies began to respond to the requests for help, more often than not in kind. Iain's sister Adine remembers becoming unofficial quartermaster (since Iain was still away much of the time at Oxford) when great parcels arrived from mysterious sources at the Shetcombe front door.

'All sorts of things turned up. I remember once a great stack of Grape Nuts being left in the front porch. Mountains of it! I just marked it up and put it in the garage with everything else.'

Of more interest to both Iain's sisters was the team that Iain and Ian drew around them for the enterprise. There were four, including Iain, from Oxford. Ian Maddick was the leader, a fair-haired medical student whose skills were to prove invaluable. Derek Springham, tall and dark, had the slightly preoccupied air of the true botanical investigator. Although he was unable to drive, which was a drawback on the expedition, he more than made up for this by his botanical inquisitiveness. David Berry, the fourth, was a no-nonsense

Lancastrian destined for a career in botany and as reliable as he was frank in his opinions.

Grafted onto the Oxford team were two more from Edinburgh to beef up the general level of expertise and, the Oxford group suspected, to ensure that all that was collected ended up where Dr Davis could enjoy it! Dr Dennis Ratcliffe was an engineer who understood Land Rovers, and put long hours in getting Land Rover number two up and running. And Salar Khan was a mature PhD student from Bangladesh (then in Pakistan). He was especially valuable as both an experienced botanist and a Muslim. He seemed to enjoy the whole expedition business nearly as much as Iain, who remembers him most for his constant smile. Salar Khan's presence on the team, his botanical experience and his gentle understanding of culture helped in many places in Turkey, a Muslim country.

The botanical search plan, drawn up by Iain after frequent telephone conversations (followed by lengthy letters) with Dr Davis at Edinburgh, was to make a virtue out of a necessity and call it a collecting expedition of 'late-flowering plants'. Most collectors would go to Turkey in the springtime, when it is especially beautiful and there are many flowers to collect. In summer and early autumn there would be fewer. But the students had no choice, they were constrained by term times and their final exam dates. What they would have to do instead was to cover a lot of ground, to try to get some idea of the overall distribution of plant species across the country and in the different climate zones.

One family they would look for was Compositae (daisy family). They would also take a wide sweep along the Black Sea coast, as plants tended to flower later there. In addition Derek wanted to collect pollen from various bogs (palynology) and salt marsh plants (halophites) from the centre of Turkey at the Great Salt Lake, Tuz Golu. All in all it was a pretty long shopping list involving a lot of driving. It was agreed that plants would be pressed sun-dried. In such a sunny climate there would be no need for kerosene-fired oven presses, such as were used in the damp and humid tropics.

The milestones on the way to the great leaving day were quickly overcome in a swirl of activity. Fixing the Land Rovers, getting sufficient sponsorship, buying equipment, sorting out visas and government permissions for collecting work. No one is allowed to go and pick others' plants without asking, though specimens of all plants collected would, as a matter of course, be sent to the herbarium at Turkey's

Ankara University. There were other, more personal, milestones closer to the day itself, including Iain's goodbyes to Anne and his mother and sisters, who were as excited as he was.

They set off for Dover in convoy, proud to be the official Oxford Scientific Expedition to Turkey. They estimated a week of fast travel out, and a week back, with ten weeks in Turkey itself. But once on the road the travelling was slow. For a start they had a major problem with weight. Although they had left behind half the gifts they had been given (including the substantial quantities of Grape Nuts) the Land Rovers were loaded nearly right down on their springs. With a full tank of petrol the short wheelbase Land Rover actually was bumping on the stops. And, despite the rebuild, it was still not performing well, and being overloaded did nothing to help.

In Belgium on the second night they made a decision to offload more stores and gear and have a complete re-pack of what was left. But where could they leave the surplus? Tentatively they approached the owner of the small *pension* where they were staying. Could he keep it for them until they returned? He thought this a great joke, chuckling delightedly as he tucked away food and equipment around his home. Yes, he would keep it all for them. And he was as good as his word. It was all there safe and sound when they returned, homeward bound, three months later.

They camped where they could on the journey, only stopping at *pensions* if they were running late on the day's journey time. It saved all the unpacking.

Although now more lightly loaded, the number two was still struggling. Just outside Salzburg it gave up. This was not at all encouraging. Three days out, half their gear left in an inn in Belgium and half their transport expired by the roadside in Austria. It seemed as if a simple climbing holiday would have been better after all. But Dennis and Iain knew this Land Rover, having had it to pieces several times back at Shetcombe. All they needed were a few tools and somewhere to work. Dennis spoke a little German. They found a sympathetic garage owner in the city itself and got down to work. By sundown it was fixed and – no doubt recognizing determined leadership when it met it – gave little or no trouble from then on. The expedition sped on its way.

One of their greatest concerns was their likely reception in the (then) hard-line communist countries of Yugoslavia and Bulgaria. Bulgaria in particular was thought likely to pose the biggest problem. They decided to stay in a hotel in Sofia to make it easier for the authorities to keep an eye on them and so avoid suspicion. This seemed to work – apart from a great deal of bureaucracy they emerged unscathed with no equipment taken away from them or undue delays imposed.

So they were slightly embarrassed to be contacted by a student who wanted to talk urgently with them and asked to be driven round Sofia. It was safer that way, he said. They were aware that to be seen deep in conversation with a young, possibly dissident, student might be seen as highly suspect. A tour of the city would be more innocent. They went, and the young man, speaking mainly in German, gave them a startlingly frank account of the problems of life under the communist regime.

'It is my duty to tell the world all that is happening here,' he said when they commented on the risk he was taking. Iain was deeply impressed with the man's courage, as they were both students of a similar age. Later, they left him and drove on out of the city, saddened by the rigid constraints imposed on one so like themselves in many ways.

But if they felt tempted to look upon Bulgaria as a sad country they were shown a bright side as well. Later, heading east, they stopped for a break deep in the countryside. Some local peasant workers came up, delighted to see them and, using universal sign language, bade them welcome in the country. After a few minutes the team made to go on. But the peasants called out to wait. They turned to their harvest crop and, picking out the best, loaded the roofs of the already heavily-laden Land Rovers with melons, smiling and nodding throughout in a gesture of goodwill. Iain was very moved that such poor people could be so generous to total strangers.

Despite indifferent maps and few signposts the expedition reached Turkey as planned, crossing the Bosporos at Istanbul. They then struck out across the north of the country. Iain took over now as scientific leader and was in his element, organizing searches and collections, noting and labelling the dried specimens and generally supervising the collection of data. He also loved camping and moving on, discovering new places, breaking new ground and the sense of adventuring with a purpose. The science gave it all direction and a sense of moral as well as physical purpose. It was good to be there.

It was a pity that Anne could not see him as he worked steadily along a Turkish hillside with his notebook and portable plant press, noting, collecting, commenting at some unexpected find or unusual occurrence.

Here, though far away from England, he was clearly very much at home. She might not have thought twice about marrying him, but as to believing him when he later promised he would go on no more expeditions . . . well, then she would know better.

At this time Iain fully expected to go on to study and research the plants that he and the others were collecting, so he felt that he was, or would be, seeing the whole taxonomic process through from start to finish. This gave added interest, and pressure, to the scientific programme. Every minute counted, every note made was important, every measurement, every sample, every pressing. For this would probably form the basis of his PhD.

Running a team too came naturally to Iain, having learned some of the qualities of leadership as head of house at Malvern College, another of Bill Wilson's legacies. His leadership style was not oppressive, simply a matter of starting out first in the morning and getting on with the next part of the programmed work in hand. His cheerful 'good morning' along with a cup of tea at the start of the day became as familiar to his team members as his quiet moments alone at the end of the day jotting notes in his diary, or sitting by his tent with his Bible open and his mind on God.

The team was not without its problems. One member, who had early declared himself particularly concerned about personal hygiene and chose to wash up and keep his own set of eating utensils throughout, became sick. Ian Maddick as medical 'expert' on the trip was unsure of the diagnosis but, talking things over with Iain, advised hospital if this were possible. They were nearing central Turkey so they took a detour to Ankara and left him there in hospital for a few days while they continued the botanical hunt round the Great Salt Lake. Iain learned from this another important lesson for the leader of future expeditions: people who expect to get ill generally do – and on balance it's best to avoid taking such people along in the first place.

But this was only a minor impediment to what was turning out to be a botanical success. They had found fewer plants than they had hoped – but this was because of the season, which they had known from the start was against them. Even so, many specimens were collected and many statistics noted. There would be much for Peter Davis to delight in.

Curiously, Iain's best memory of this, his first botanical expedition, was not of the plants, but of people. The team members and how they worked together, and also the people they met along the way, in particular the people of the mountains. During their first weeks in Turkey, while climbing in the Lycian Taurus mountains, they came across Kurdish herdsmen and women watching over their goats in the high summer pastures. They lived in goatskin huts and ate flat dry bread and yoghurt – a semi-tribal, semi-nomadic society, as far removed from the sophistication of the university expedition with its cameras and plant presses and off-road vehicles and calculated diet as it was possible to be. But Iain (and Ian Maddick too) took to them straight away.

Their interest was reciprocated. They shared the nomads' hard bread and told them, as best they could in sign language, the purpose of their visit. Their simple welcome and understanding struck a chord with Iain, as had the warm welcome of the peasants in Bulgaria. He found he responded to the basic feeling for the earth which he sensed in such people, a feeling that seems to have been abandoned or forgotten in the West. There was so much to learn from people like these. Iain is not now, nor was he then, under any illusions about noble savagery or the joy of poverty; he simply believes that in becoming 'civilized' we have lost, as well as gained, much, and that a close relationship with the earth is one of those things. This, he feels, is perhaps why so many educated people are environmentally irresponsible.

'These people knew what it was to rely on the earth,' he says, 'and that found an echo within my soul.' Later, much of Iain's work among the Amerindians was to reflect and reinforce this feeling.

His other abiding memory is of the grandeur of the mountains. One morning they climbed to the top of the Ak Daglari (White Mountains), part of the Lycian Taurus range. Iain looked out at the view spread out before him, and – for once – did not quickly focus in on a nearby plant to note its physiology or family. There were few to see at that altitude anyway, but more than this he was struck, almost overcome, by a sudden sense of awe. Awe at the massiveness and beauty of the scene, for him the creative handiwork of God. Someone had been there and planned it all. He felt privileged to see it, and found himself thanking God for it. He has felt this occasionally since, when he has glimpsed a jungle vista, though it is less common working under the dense canopy of the rainforest. But even there the massive cathedral of tree trunks in primary forest is awe-inspiring. And awe, he feels, is a good thing, partly because creation is something to be held in respect, and partly because it takes away the notion that nature is a toy for humankind to play with, pick up and put down on a whim.

A conservationist, believer or not, cannot think in that way, but must have a genuine respect for the natural world before he or she can work consistently to save it. 'But,' he says, 'for those of us who are privileged to see such sights that respect comes very quickly.'

As autumn drew on, the expedition made its way north to the Black Sea coast then finally west again to Istanbul, and home across Europe. On the expedition Iain learned that he loved field work more than even he had expected. But his high hopes to go on to research the collected plants in Edinburgh for a PhD came to nothing. He was not selected for a place.

In Oxford Frank White had been watching it all quietly. He already knew Iain's capabilities from the summer vacation work he had done for the Forest Herbarium in Oxford. Now he had seen him put an expedition together, one that had worked well. He knew that Iain could do a practical as well as theoretical job. There was a research assistant post available for a month or so in the Forestry Department. Perhaps, he suggested, Iain might consider that? He could start his DPhil studies there. In this rather informal way Frank White became Iain's supervisor and he commenced his DPhil work. He would have to wait three years for his next expedition, and although he was very busy, to him it seemed a long time coming.

6. NEW WORLD

The Chrysobalanaceae were, so to speak, at the root of Iain's next move. His thesis, entitled 'A Taxonomic Study of Chrysobalanaceae', was as thorough a piece of plant classification as it was innovative. Despite the unorthodox use of computers, his effective integration of wood anatomy, seed morphology and detailed observations led the examiners to award Iain his Doctorate of Philosophy in Forest Botany. It was a good piece of research.

But where, if anywhere, would it now lead? Once again there was no clear path ahead. In many ways Iain was back at the start. The next stage should be 'post-doctorate' work – yet another learning ladder. Despite six years of degree-level botany behind them, 'post-docs' were still very much just research assistants, still on probation as scientists and having to prove themselves. But that was the usual way forward.

At this point Iain's earlier contacts came to his aid. In researching the Chrysobalanaceae he'd written to many of the major herbaria around the world to request specimens. These included Kew in London, botanical centres in Utrecht, Paris and Munich on the Continent, and Washington and New York in the United States. The New York Botanical Garden (NYBG) in particular was very helpful. They proved to have a strong interest in South America, and the Chrysobalanaceae in South America were in great need of a revised classification. New York lent Iain a number of specimens.

There were, however, a fair number they did not give him. They had a particular collection of Chrysobalanaceae which the head curator Dr Bassett Maguire had collected personally and he had reservations about firing them off across the Atlantic to the first DPhil enthusiast with a research interest. On the other hand, he knew he did not have time to research them properly himself, much as he would have liked to have done. Botanical research is like that. There is so much to collect and such an urgent need to do so – especially recently as environmental crises seem to be striking so many of the unstudied areas of the globe – that expedition members usually collect much more than they can ever hope to study. Also, of course, many collect plants which are not really their own speciality, hoping that someone will make use of them in due course. But if the curator was unwilling to send his whole collection to England, he was well placed to consider recruiting someone to do some work on it in New York, where he could keep an eye on progress. There was a small snag. Even with the right references, whoever it was would be required first to join an expedition in South America, in Surinam. The New York Botanical Garden could use the help in the field and there were a lot more Chrysobalanaceae to collect.

Iain, of course, saw the requirement to go on an expedition not as an obstacle but as the chance of a lifetime. It was just the type of botany he liked and an excellent start to his scientific career. It also seemed to be the only real option for post-doc work open to him. The project in Africa, suggested by Frank White, didn't feel comfortable (in fact it folded within a few years) whereas to work with the Chrysobalanaceae would be enjoyable and give him a real boost in making his mark. Anyway, it was good science. He knew so much about this graceful family of tropical trees already.

These and others were the arguments he had rehearsed before Anne in their caravan, as she grew larger with Rachel and poured over the graphs and charts he had assembled for his DPhil thesis. Anne was not terribly impressed with the idea of her husband leaving her at just about the time their first baby was due, though naturally she was pleased he had found a way to forward his career.

The idea of moving to New York was both exciting and forbidding at the same time, though naturally Anne had some idea of the place from films, and the city didn't have the friendliest reputation. The plan was for her to cross the Atlantic, by steamer, with the baby, to join Iain when he returned from Surinam. He should be sure of his position in the NYBG by then. In the meantime Margaret Prance would have Anne and the baby to stay at Shetcombe. They could sell the caravan to help fund the transatlantic move.

So it was agreed. It proved to be the first of many 'botanical separations' for the young family. But they felt that God was with them in this decision, and they could trust him with the consequences. That mattered to them a great deal.

So on 24 August 1963 Iain took a short flight to Amsterdam, changed planes and then flew onward by KLM DC 8 via Lisbon direct to Paramaribo. Capital of Surinam, it is on the north-east coast of South America, the uppermost edge of the Amazon rainforest.

Surinam had recently gained independence from Dutch colonial rule and to Iain's eye, as he stepped out into the blast of tropical heat after a day in the air, it was as if he was back in Holland. The Dutch influence was everywhere. The city seemed made totally of wood, with tall weather-boarded buildings in the Dutch colonial style lining the streets as hotels, apartments and private houses. There was even a wooden cathedral!

It was very humid and he felt he could sense the strangeness, the 'differentness', of the tropical vegetation all around him, although he could see little of it in the darkness, just glimpses of it in the headlights of the taxi as they rushed from airport to city late that night. All the nationalities of the world seemed to be in Paramaribo and the noise (right through the night) and heat of this cosmopolitan city made him fear he would never manage to get to sleep. He had been told on the plane that the native language, Surinamese, was a mixture of Spanish, English and African, with some Dutch thrown in, and that the people were similar. Now he believed it. For all the noise and excitement, exhausted by the journey he went straight to sleep.

The next day, keen to move inland, he had to wait for the complicated expeditionary 'field train' arrangements to fall into place. As these included flights, canoe trips and an overland march in stages with local guides, the whole thing took several days to organize. Waiting for arrangements to fall into place on botanical expeditions is as common as plant collecting, especially in South America. Western impatience quickly learns what can and cannot be done. In the meantime Dr Jop Schulz, from the Surinam Forestry Department, played host to Iain and he was able to offer a special treat.

He took Iain a little way out of the city to a white sand savannah area and showed him his first living Chrysobalanaceae, *Licania incana*. It was one of the less imposing varieties, no more than a large shrub, with soft green leaves and small white flowers. But there it was, growing in abundance on open land near the airport. For Iain to see a plant he knew so intimately, but until now had seen only as monochrome piles of dry brown pressings, was a moment to be savoured.

In about four days, arrangements had been made for him to travel to the interior and he boarded a twin-engined Beechcraft twelve-seater plane for his flight inland; stage one of his journey. The pilot was in a sight-seeing mood and took them over a Brazilian Amerindian village of the Arawak people *en route*. It was Iain's first sight of one of the peoples he later came to know well. Then they flew round the Toemahac Hoemahac and Kassikassima mountain ranges, the latter with its distinctive twelve peaks thrusting up high out of the green jungle. Iain was entranced by the beauty of them and wanted to land and start collecting immediately on their slopes. Mountains have a particular appeal to botanists since plants and plant populations change so much with quite modest differences in height (and the corresponding change in temperature). On a mountainside collections can be made from several different environments on one climb. Provided, of course, that you like climbing and don't mind carrying many extra pounds of limp vegetation on your descent. Botanizing at high altitudes demands a robust constitution. But this time the plane flew on.

They landed finally at Kayser airstrip, cut out of the jungle on the banks of the river of the same name. There Iain was met by his first contact from the New York Botanical Garden: Noel Holmgren, a PhD student from Utah in middle America.

Noel, though no older than Iain, was already a man of considerable jungle experience, having taken part in the first weeks (phase one) of the expedition. Determined to make an impressive start, Noel took Iain off right away, hacking into the forest, for a short walk to a nearby area of savannah. Things did not go well. Without warning, Noel embedded his machete in a hornets' nest and the angry occupants took quick revenge stinging him harshly over one eye. With one eye completely closed, he continued his guided tour into the jungle. As they went he fielded a flow of questions from Iain about the jungle flora, the characters on the expedition, and the likely lifestyle of a post-doc at the New York Botanical Garden. Iain was very much aware that he was on probation on the expedition and keen to find out how the land lay so that he could make a good first impression. He had already taken the drastic decision to shave off the distinguished-looking bushy beard he had begun to sport at Oxford, not wishing to appear unkempt and sloppy to the Americans. As they walked further and further into the jungle the conversation continued. Noel remembers:

'After about an hour I was completely lost. I had no idea where we were. But I sure wasn't about to say anything to my new British colleague, who was hanging on my every word. It was good being the local expert for once. So I kept turning left, hoping to strike a familiar path or some track. No such luck. Dusk was drawing on and I could sense in the conversation that Iain was edging round to suggest we drew this "ramble" to a close, something I had been trying to do for the last half hour! So it seemed I ought to come clean. I opened my mouth to give him the biggest shock of his trip so far, when suddenly, away in the distance, came the rumble of a diesel engine. Someone had started up the airstrip generator. "Time to be getting back," I said. "That's the airport power plant." I headed towards the sound, soon found a track and we hurried on back to camp. He never knew a thing about it until I plucked up the courage to tell him later. We might have been lost for days, or worse!'

Iain later made a personal resolution always to carry a compass when wandering off on walks in the jungle, no matter how expert the guide. 'At least then you can work your way down on a bearing line to a riverbank. And if you know which way the river is flowing you have some chance of getting back to camp. You may also be spotted. In the rainforest the rivers are the highways and your chances of rescue are much greater.'

The next stage of the journey was along these watery highways, with Noel and more expedition supplies, by dug-out canoe. This was to be followed by a final eight mile trek through the jungle to the latest field camp. This was 'Camp $14^{1}/_{2}$', where the current collecting work was based. It got its name from the distance in kilometres along the 'transect' or paced out research line driven into the jungle by the expedition leader Dr Howard Irwin. Camp $14^{1}/_{2}$ was likely to be drier than the base camp, Camp Lucie (named after the local river) which was on the riverbank and had been flooded out twice since the expedition's arrival.

The three dug-out canoes in which they travelled were manned by the Djuka people – not Amerindians but descendants of escaped slaves who had re-formed something of their African tribal culture in South America. The Djukas staffed this expedition as cooks, carriers, boatmen and forest huntsmen, being expert at all types of jungle living. Two crewed each canoe; one worked the outboard motor slung over the stern while the other stood in the prow to fend off rocks and other passing dangers with a long pole.

They spent two days travelling in the dugout, running swiftly down the Kayser river, staying overnight at Falls Camp (another interim base) and then forcing their way upstream on the Zuid river the next day.

The camp food at each stop, Iain recalls, was a strange mixture of the local staple, cassava (which needed washing down with 'a lot of coffee'), expedition basics (breakfast one morning was porridge with corned beef) and intriguing luxuries from the forest – venison and iguana killed and cooked up by the Djukas.

On both rivers there was a lot of work for the pole-man to do as shooting rapids was the norm rather than the exception. Iain coolly described this as 'most exciting' and later noted that he saw his 'first macaw, iguana and alligator'.

Camp Lucie was reached unscathed and they landed and spent the night there. There was no danger of flooding since, with the onset of the dry season, the water was receding rapidly. Noel mentioned this to Dr Irwin when they caught up with him, in case it called for a move. For them all (including the field party) to be left stranded up-river was much more serious than occasionally being flooded.

Next day they trekked out to Camp 14½, some seven hours of jungle hiking. It was a tired and footsore pair of junior field researchers who finally met Howard Irwin and Tom Soderstrom, another young American botany student, at the field campsite that evening. Dr Irwin (who wore a beard!) put Iain quickly to work, blisters and all, pressing the plants already collected that day. Iain was delighted.

He was hugely impressed with the camp and camp life. This was an expedition in the grand style, with teams of Djukas at each base camp to help run the site and equipment. There was a transportable field laboratory, four 'white gas' oven presses (Iain never discovered what 'white gas' was) and even a communications headquarters complete with pedal-operated HF radio transmitter, in the charge of one Sergeant Piper from the Surinam Sapper Corps. Every day on his signals 'scheds' he would tap out, in Morse code, the requested movements of expedition stores, aircraft, mail, pressed specimens, petrol, food, dug-outs and all the other items variously distributed up- and down-river to keep the expedition supplied. All in all, it was an impressive operation.

The expedition members lived in tents or open-sided thatched bivouacs (tents could become very claustrophobic in the humid tropics) though they wrapped up well at night as it was cold. They slept in hammocks, which was essential; sleeping on the ground invited an army of irritating insects, ticks and small animals, not to mention one or two deadly ones, to come in with you for warmth. No one likes to wake up with deadly snakes or scorpions insisting on their half of the blanket. A hammock can also easily be enclosed in a mosquito net.

Similar shelters were provided (back at Lucie Camp) for the field laboratory and the pressing ovens. Here too the warmth encouraged occasional unwelcome visitors. Early on Iain disturbed a very large scorpion dozing by the hot presses one morning. It rapidly ran off before it had the chance to sting anyone.

Bathing was in a nearby stream, where a group of rocks formed a natural bathing-pool. A scrub down usually brought the busy day of collecting to a close, but only after the day's collected plants had been put in the presses and those just pressed lifted out and packed up for dispatch down-river to the airstrip. After supper Iain would chat to the others, write his diary, or a letter to Anne, his mother or Frank White back in Oxford. Then he would read his Bible by the light of a storm lantern. Evenings were short, for it was always an early call at five thirty or six in the morning for the botanists, who needed to make use of all the available daylight for collecting. And sleep was sound, except occasionally when the nights were full of drumming.

The Djuka field helpers particularly liked to show off their skills in carving, usually of boat paddles and drums, the drawback being that they then enjoyed demonstrating the drums through the night. One night Iain records being much disturbed by a furious argument among the Djukas as to the number of full moons which had occurred so far that year. The botanical team were eventually appealed to as, rather weary, arbiters on the matter.

The other unofficial members of the expedition were the pets, which seemed to be acquired almost by accident as time went on. Macaws were the most popular, followed by the more intrusive monkeys – one ran up Iain's arm to join him in eating supper! These pets were 'expedition only', at least for the botanists, and not taken back to America. At jungle camp Iain was in his element and his diary entry for 1 September 1963 records with great satisfaction: 'At last out collecting.'

Most days were spent in collecting, along the line of the transect, up and down the riverbanks, or on rocky outcrops of one of the Wilhelmina mountain range which was near Camp 14⁻. But they were soon on the move, as the lowering water level had forced Dr Irwin to make an early departure back to Camp Lucie. Neverthless collecting continued all the while.

Shortly after Iain's arrival there was an odd incident, during collecting, which has stuck in his mind. It confirmed his thinking about the quite different view that a westerner (and botanist) may have of plants compared with that of a local of the forest.

Frederick, a Djuka, was the camp cook and as such wholly familiar with the running of expeditions, particularly botanical ones. He would always volunteer, in fact. In those days one of the standard methods of collecting specimens from tree species was to fell the tree. The flowers, fruit and leaves were almost certain to be at the top, at least twenty metres up, and few expeditions then had modern climbing equipment or many trained tree-climbers to get into the rainforest canopy (things are very different today).

Iain had spotted a Chrysobalanaceae (*Licania*) near the base camp and decided to fell it, calling Frederick to help him. But instead of the willing, experienced help he expected he got a very definite no. Surprised, he asked what the problem was. 'The Bushy Mama tree spirit,' replied the cook. 'She would not like the tree cut down.' He would not go against her wishes, not for anyone. Iain was adamant he wanted the tree collected. Frederick pleaded, but Iain was determined. Eventually Frederick saw he would have to agree, but asked for time to explain to the Bushy Mama what was intended and at whose door she should lay the blame. The explanation took some time, and afterwards, as he swung the axe deep into the bark, Frederick continued to explain pitifully to the tree that he was under orders and did what he did because he had no choice in the matter. Iain was no believer in Bushy Mamas, but he was nevertheless challenged by what had occurred not only to respect the forest as far as possible, but also to take into account the relationships others may have with it. Others, particularly indigenous peoples, have a stake in the forest which can be quite beyond the scope of western comprehension. Iain also began to see that felling might not be the best way to collect. He felt he would do well to remember that, if he was going to spend time here in the jungle. He and Frederick remain the best of friends.

This was a lesson Iain certainly took to heart, for his work on the relationship between tribal cultures and the rainforest not only later ensured the success of a whole new department of botany (called ethnobotany) at the New York Botanical Garden, but also provided the basis for much modern thinking on world ecology and the sustainable use of natural resources.

The expedition drew to a close at the end of November and Iain and the rest of the party returned to New York. By that time it had been tacitly agreed that Iain was the sort of researcher the NYBG would like to have working on their Chrysobalanaceae collection and they asked him to start right away.

Anne was contacted at Toddington and told to pack. Iain scoured the New York area with Howard Irwin, looking for a cheap apartment. They found one to the north of the city in a suburb called Hartsdale, in Westchester County, a dormitory area of New York City. At the time they thought Iain and Anne would be there one or possibly two years at the most. But although Iain did not know it then, this area was to be their home base for the next twenty-four years.

Anne boarded the *SS Queen Elizabeth* in October and took baby Rachel on deck to bid Southampton, and her mother-in-law and sister-in-law, farewell. Four days later she went on deck again to see the Statue of Liberty slide past, excited at the prospect of soon seeing Iain. She also realized that she and her husband and baby, though welcome, were, like so many before them, coming very much as strangers into the New World.

7. HOME FROM HOME

The garden apartment which Iain had found was pleasant and modern and ideal for a young family. He could commute easily to the New York Botanical Garden either on the subway train or straight down the Bronx River Parkway in their tiny car, a very British Mini Cooper. The gardens, although situated in the Bronx (not the brightest part of the city), were a reasonably secure haven of peace and green. They contained the only remaining original (hemlock) forest from the local area, some forty acres of it, within the grounds of the gardens themselves.

Iain soon settled into the Science Department, quickly learning his way about the herbarium. Anne settled into American suburban life, first as the 'new English mom with the cute baby', and then as one half of the 'friendly church-going couple from the Botanical Gardens'. They had decided to attend the grandly-named First Baptist Church of White Plains, which, although it wasn't the sort of church the Prances had been used to in Witney, well suited their active belief and relatively informal and personal expression of faith. Anne recalls that they decided on this church mainly because it advertised its services on the entertainments page of the local paper – a sign, they felt, of spiritual energy and community involvement.

But it was not to last. For Anne had one more role to adopt – one she had been promised would not be asked of her – that of botanical expedition 'widow'. After seven months of life as a family unit in Hartsdale, Iain sheepishly explained that his research was going so well that he had been asked to join another tropical expedition. This time it was to Brazil, driving north from Brasília in the centre of Brazil to the eastern part of the Amazon basin. He'd be away about four months, he thought. Not too bad. And, with little more comment than that, rather like the Cheshire cat in *Alice in Wonderland*, he faded gently from her sight leaving the ghost of his boyish grin behind.

Then, for Anne, America became a very lonely place indeed. Not that people were stand-offish. Many went out of their way to be kind and friendly but Anne, who was living off little more than a student's income in what was a very affluent suburban area felt that no one really understood her situation. To be married to a roving botanist was considered pretty odd by her neighbours anyway, without throwing in the fact that she was an English country girl in the city suburbs and in the United States for the first time. A great deal of their money was going on the apartment – which had to be in a 'safe' suburban area – so she did not feel she could go out very much (she would not spare any money for a baby-sitter) except to church on Sundays. The church people were pleased to see her too, but were generally unaware of the pressures upon her (the 'culture shock') as a stranger abroad.

In addition, many people half suspected the Prances were actually having marriage problems – why else would he have left her like this? – so they maintained a tactful, cool reserve toward her. There was the occasional formal invitation to lunch or morning coffee, an hour or two of Hartsdale or White Plains local chit-chat, which, being an outsider, she could neither add to nor dispute. Then so often it was back to her apartment – alone with baby Rachel.

She had no family to turn to, no real friends to call on when she'd had enough of the baby, who missed her daddy badly. No one who would drop in on her. No one had really had time to get to know her, or appreciate how lost and disabled she felt amongst strangers and with a small baby.

Sundays were the worst. Self-pity took over. She became convinced everyone was going back home after church to happy gatherings and good family times together while she alone in New York had no husband around and just a meal for one. Added to this, she did not hear from Iain in Brazil for eleven weeks, which worried her a great deal. Iain's expedition in this case was unusually fast-moving and the rear link was slow to connect and bring mail out back to the USA.

She says, 'For me it was hell. Someone would ask me in church if I had heard from Iain and I would just burst into tears. And they would take that the wrong way!'

Just one evening stood out, when an American friend kindly sent in a baby-sitter to look after Rachel, and took Anne into New York City to the theatre. It was her friend's way of giving her a birthday treat and it was such a bright spot in all those blank months that she determined never to get herself in that position again.

She kept to her decision. When Iain returned she told him point blank that the next time he went off she was coming too, wherever it was, and bringing the baby! She knew it was likely to be the Amazon, but she didn't care. By that stage, anything had to be better than living in suspended animation in New York.

So, in the summer of 1965, when Iain went south once again, Anne, Rachel and their new baby Sarah went along too. Sarah was just ten weeks old at the time and her appearance on the scene had seriously threatened to jeopardize Anne's plans in the preceding months. But, curiously, it was the family paediatrician who encouraged them to go ahead with the idea.

'It's best if the family stays together as much as possible. Better for mother, better for baby, emotionally. And that is very important.'

They took heart from his advice, for most of the people they knew in Hartsdale, and the Botanical Garden for that matter, thought she was mad. Anne herself often veered towards the same opinion when she looked at the undertaking, though she repeatedly reassured herself with the thought that nothing could be as bad as staying behind.

In late July they packed and, letting their New York apartment, flew south to Florida and south again across the Caribbean to Belém in Brazil at the mouth of the Amazon. From Belém they flew on inland to Manaus, the city which lies at the heart of the 2.5 million square mile Amazon basin. Then a city of a quarter of a million people, Manaus is 1,100 miles inland from the Atlantic coast, built on the banks of the largest and probably the most famous river in the world. So large that the biggest ocean-going cargo ships and steamers can comfortably make the journey from the sea to the city port up-river and beyond. The Amazon continues inland for more than twice that distance, flowing west to east across the widest part of South America from its source in the snow-capped Andes of Peru only 120 miles from the Pacific coast.

Manaus grew up on the rubber trade in which Brazil dominated the world at the turn of the twentieth century. Many rich merchants who controlled the thousands of acres of rubber tree forests (plantations of rubber trees alone are not possible in the Amazon) settled in the inhospitable rainforest climate and built large houses on profits made from the latex collected by the rubber tappers from trees up and down the banks of the mighty river. Manaus even boasts an elegant opera house from that time, now one of the tourist attractions of the city, for rubber brought all that the Americas and Europe regarded as civilization and culture to this heart of the jungle.

But Manaus was still an island in the rainforest. There were no roads built to the city, even by the time the Prances moved there; the only way in was by plane or ship. And by 1965 Manaus had lost much of its *raison d'être*. In the twenties it was found that rubber would grow well, on plantations, in Malaya. It was taken there by Kew botanists to ensure a wholly British colonial supply of this vital commodity. And the automatic source of revenue the Brazilian rubber magnates had come to rely on began to dry up, like an exhausted rubber tree. There was still income, but not at the same high prices. Manaus went into slow decline and that is how the Prances found it in 1965. Anne says that it seemed 'decayed and dormant'.

But the state of the city and its faded splendour were the very least of their problems. They were met at the airport by a missionary family, Lonnie and Janelle Doyle, who lived in Manaus with four of their five children. They greeted them warmly, and Anne was astonished to be given as a 'welcome gift' a bottle of water. She thought this a little strange, though she made no comment. But she quickly realized its importance.

All water drunk in Manaus had to be thoroughly boiled, then cooled down for children – especially for baby feed bottles. Anne became suddenly grateful for their foresight – it would have been impossible for her to have given her thirsty and hungry children what they needed in the tropical heat without that gift. But it wasn't long before the problem of boiling all their water seemed rather trivial compared to some of their other living problems.

Their one-room flat on the tenth floor of a harbour-side apartment block (the only high-rise building in Manaus at the time) was certainly large enough, but the elevator only worked when there was power available in the city. As this supply was somewhat random, it was not to be trusted. Anne soon learned that carrying two babies and all her shopping up ten flights of stairs could keep her fitter than a mountain botanist. The kitchen was small with an elderly cooker (which Iain had found as the apartment was unfurnished) in the corner, fuelled by bottle gas. The door would not stay closed without the assistance of Scotch tape, but was fine when stuck up.

The bathroom (shower and toilet) was not fine. This was largely because the other inhabitants were two-and-a-half inch cockroaches. Cockroach eggs festooned the walls and door and no campaign to exterminate them, and there were many, seemed to do more than reduce the numbers temporarily. In the end, they all adopted a 'see how thin you can be in the shower' policy, so that they touched nothing around them while enjoying the modest dribble of water.

The food they ate was, naturally, Brazilian. In later years, when the city was declared a 'free port' (which did much for the central Amazonian economy) more foreign food, mostly American, was imported. Brazilian staples are rice, beans and manioc. Added to this was fish from the river, some meat, and fruit and vegetables in abundance. Anne found that she could shop for these at an open-air market, though this moved to a different location in the city every day! The problem was the quality and hygiene. It became her habit, as it was with the locals, to handle everything very thoroughly before buying to test its freshness and quality. Even so a bowl of rice had to be picked over grain by grain back at their apartment to ensure she did not cook small stones, river mud or insects (which were always there in quantity) in with the rest of the dish.

Meat was bought in gory lumps, several kilos at a time, having been hacked off a warm carcass in the market place to order, usually on a dirty wooden bench, by a very sweaty butcher. Then it had to be washed and re-butchered by Anne at home before it could be used in the week's meals. Fruit and vegetables (of which there is an amazing variety in the Amazon: okra, carira – a tropical spinach – graviola, pineapple, papaya) and nuts, having been checked over thoroughly by Anne (and a dozen other shoppers beforehand), were taken home, washed and then soaked in a strong disinfectant – iodine – solution before eating.

Not only had all water for drinking to be boiled, but all the washing-up – plates, saucepans, cups – had also to be rinsed in boiling water as part of the basic hygiene. To miss out any of these tedious stages, which consumed much of the day in 'simply living', meant a real risk of stomach illness and, more sinister, of ingesting tropical parasites. And Anne had to be doubly sure because of the children.

Washing clothes, especially with a new baby, was a priority. Not long after they moved in, Anne opened the door to a little old lady who offered her standard washing service to new tenants. Anne was delighted and was just about to offer her the whole dirty pile to deal with when she was struck by the thought that she did not know when she would get it back. The lady said it would be returned, 'the day after the sun shines'. An awkward discussion ensued, frustrated by the language barrier which forced Anne to refer frequently to her Portuguese dictionary. After some minutes, during which the little old lady was clearly keen to get on with the work already offered, Anne began to understand properly the flexible nature of the washing arrangements. The whole business (in this rainforest climate) depended on the sun. When it shone the washing could be turned round in short order. When it didn't, there could be long wait. Anne quickly divided her pile into two: 'urgent' and 'not so urgent', and handed over the latter. It came back, beautifully washed and ironed, a week later. Not too bad as it turned out, but she was glad she had kept the first pile. Hand-washed, it had kept the family moderately well turned out until the other came back.

Despite moving to the Amazon, it was obvious that Anne and the children could not actually go on collecting expeditions with Iain. That was a practical impossibility. So in a sense Anne had exchanged one sort of isolation for another. 'But at least,' she says, 'I was in the same country!' And as the expeditions were, more often than not, a series of forays into the interior from their base at Manaus, she did actually get to see Iain at least on a month to month basis, if she was lucky.

Nevertheless when he left in August to go out collecting again, after not much more than three weeks together in their new high-rise home, time hung on Anne's hands. But she found that the dockside location of their room gave her unexpected entertainment, which she has never forgotten.

'The river view was a constant source of interest, and as we spent many hours watching all the movement around the docks I learned just how important the river was to the city,' she recalls.

Their windows looked down on the dock area where the large ocean-going steamers tied up and smaller craft plied their subsequent trade to out-stations up- and down-river. The main dockside was a massive floating pontoon system, built to accommodate cargo ships and tankers of many thousands of tons, so that they could rise and fall according to the annual fifteen-metre change in water level from rainy (flood) to dry season. In between these large vessels, the smaller riverboats or tiny canoes paddled or motored in and out carrying anything from piles of oranges, fish and vegetables to stalks of bananas and exotic fruits.

Anne graphically recalls the typical early morning scene when, perhaps awakened before dawn by a demanding baby, she would gaze out of the window to drink in the quiet of the sleeping city and the dark swirling river before most people were up and about.

'There is an hour just before daylight when the tropical sky is washed with a strange water-colour light. As the last star holds onto night the silence is profound. Just as the real daylight comes the silence is broken by the throb of a small motor on a canoe. It is a market vendor hurrying his wares down-river to secure a good selling point at the rivers-edge market. It's as if he holds the keys of dawn – after he passes the city opens to a new day.

'This one room was my home for about five weeks. Primitive and difficult though it was I shall never forget nor regret the time we spent there because through the windows I got a feel of the city which was to be home to us, for alternate years, for some time to come.'

In a month or so Anne was able to find a stone house to rent which offered them more room, neighbours (whose children immediately taught Rachel to converse fluently in Portuguese, while Anne was still struggling with the dictionary) and no arduous climb on her return from market.

There they acquired on loan a fridge, big, old and green, which worked on kerosene, kept in a big fuel tank at the bottom. 'The hotter you got the burning wick underneath, the colder the things kept inside,' Anne remembers, 'and if the wick wouldn't light a good kick would usually get it going!'

Their furniture (the house was unfurnished) was mainly camp equipment too old and battered to be taken on expedition. The worst trap for an unwary visitor was the dinner table, an ancient relic of much earlier field trips made by long-forgotten tropical explorers. If leant on too much it would suddenly give way and collapse backward over the offender, delivering all the meals and full table setting into his or her lap!

There were no beds, hammocks being first choice. Not only because of the cockroaches (and other things) which walked abroad at night, but also because hammocks were cooler than beds and don't need making up every morning. All the houses were fitted with hooks in the walls to take them. They are a common sight up and down the Amazon where even guest-houses provide 'hooks and breakfast' rather than the familiar 'bed and breakfast' of western taste. Even the large passenger ferries which ply the thousands of miles up- and down-river have hooks for the passengers' comfort.

Anne remembers the first shock of introduction to a culture and a city which she has not exactly learned to love, but which has become very much part of her, part of her family and part of her personal insight into the problems of adapting to a new culture. She has not forgotten her feelings in New York that first year, either.

'There is no one quite as helpless as someone lost in a foreign culture,' she says. Later, in New York and Manaus, she used her skills as a teacher of English to help many who found themselves in this baffling and alarming situation, not only by teaching them her language but also easing their passage into the new culture. Wives of Japanese diplomats and businessmen appointed to the USA, Brazilian botany students trying to come to terms with academic college demands, and those at Kew today, who may be thrust into an important sponsor's formal dinner party and are clearly not 'at home', all these have had cause to thank Anne for her special understanding and empathetic help.

And what of Iain in those days? Around the time of Anne's move from apartment to house his diary for the week laconically records the usual steady catalogue of new plants collected and odd occurrences from the wild interior:

'Woken suddenly in the night by a shot which made me jump. One of the men came back with a five foot alligator, from the tiny stream by the camp where we bathe. It was amazing to see such a big one in so small a stream.

'Made a good collection today of 42 numbers out to Kilometre 8.5, but feel we are all still pretty unfit as took a long time to cross rough country.' (A 'number' is one plant collected, there may be up to 12 specimens of the same plant for one number recorded. All must be carried back to the base camp for pressing.)

'I have never seen so many snakes as in this area. Today we came across two poisonous ones along the line.

'Last night we had two sick with fever, shouting, etc., all night, so we had very little sleep at all. We are just hoping this fever will not spread any further. Bento bitten by a large bush spider.

'The weather is remarkable.'

At least, as Anne said, they were in the same country.

8. AMAZON CURATOR

Over the next ten years Iain and his family split their time between New York and Manaus. When Iain was in the field and Anne and the children in Manaus, Anne naturally became the rear link for many of the expeditions – rather as Adine and her mother had done for Iain's first Oxford expedition from Shetcombe.

Through the stone house in Manaus parcels and personnel designated 'for the next trip with Iain', would pass and Anne would sort out visas and customs permissions for them all, write off for new equipment, dig out maps, book light aircraft flights to the interior and buy in supplies.

'In fact,' she says, 'I got too smart for my own good and instead of Iain returning from one expedition to spend two or three weeks at home while he sorted out the next, I had it all wrapped up for him. So he was back out again in three days. Not very bright!'

Anne's willingness to set up home and organize things for him in Brazil was not the only reason why Iain was able to devote so much time to active Amazon exploration. Since he was officially on the research staff of the New York Botanical Garden he would normally have been expected to spend much more time in New York than he did. That he could commit himself so much to the Amazon was largely thanks to one man: Dr B.A. Krukoff.

Although Iain had started at the NYBG through an expedition, and the institution very much encouraged these, all such field work had to be funded by a grant from an outside source such as the National Science Foundation, a US government agency that sponsored scientific enterprise throughout the country. This funding was always tied to an individual project, so each expedition needed full justification on its own merits. It was more difficult to raise funds for a programme of visits over a number of years as his employer, the NYBG, would always have first call on his services in New York. They had many ongoing projects to which his contribution would have been most welcome at any time.

On the other hand Iain's situation was helped by having made contact with (and a good impression on) the Brazilian government, so some of his expedition expenses were already being met jointly by both the USA and Brazil, something which was to prove useful later. Brazilian botanists were setting up their own national programmes and at that time – in the mid-sixties – both governments were anxious to co-operate in research and exploitation of the Amazon region.

This interest was not sinister. Both genuinely wanted to encourage development in the Amazon, seeing it as a very 'backward' region. But it was done in the wrong way. This was largely due to a serious lack of understanding as to the likely environmental and anthropological consequences of the plans they put in train for the region. In those early days Iain himself was not fully aware of the possible environmental impact of the various development projects he heard about on his Brazilian travels. For one thing he was much more interested in 'just doing botany'. And he felt, as many did, that the opening up of the Amazonian interior, especially the building of the Trans-Amazonian Highway, would enable him to find and classify much more quickly the estimated tens of thousands of uncollected species of the forest. Later this perception was to change radically.

Iain had already become involved with 'INPA' – the Brazilian National Amazonian Research Institute (Instituto Nacional de Pesquisas da Amazônia). He had involved their staff on some of his expeditions and the New York Botanical Garden were keen to further this link and make a thorough going appraisal of work on Amazonian (and tropical) plants in general. The work of Drs Bassett Maguire and Howard Irwin needed a sound follow-up. Iain looked very much like their man for this – but where was the money? They knew he would need to find a benefactor who felt as strongly about this part of the world as he did, and who cared enough to pay him to be there.

Iain first met Dr Krukoff in the herbarium at the New York Botanical Garden one cold winter's day in 1964. Krukoff was a White Russian *emigré* who had come to America penniless in the twenties and made himself a millionaire through botanical speculation. He was not a man to be crossed, though Iain had not been warned of this at the time. Krukoff was not only wealthy – a fact which gave him instant status in America – he was also a botanist of genuine wisdom and experience (the latter hard won through a career in the tropics). Krukoff felt he had seen too much of life, both botanical and human, to mince words with any mere NYBG employee. His gruff voice, still strongly accented, could usually be heard several corridors away in the NYBG as he drew some poor unfortunate's attention to a less than perfect piece of taxonomic observation, or the omission of a vital element of data.

The arrival of Krukoff's chauffeur-driven limousine at the gates of the NYBG (a weekly occurrence in the sixties) invariably caused shockwaves of anxiety around the institution. From the vice-president downwards, the staff braced themselves to meet the inevitable repeated demands for plant specimens, slides, notebooks, microscopes and monographs. For Krukoff nothing was ever done fast enough!

It was perhaps inevitable, given the strong personalities of both men, that Iain and Krukoff should clash. But it was not any failure on Iain's part that provoked the great man's wrath: it was Iain's fresh way of looking at things. Still very much the new boy, Iain had been asked to explain his work on the Chrysobalanaceae. Doctor Krukoff indicated that he'd had an interest in the plants from the Amazon region for many years. This was an understatement. Iain learned later that he had not only explored the region extensively, but had also bought a plantation in Guatemala just before the Second World War. It grew *Cinchona*, the natural source of quinine. When the Japanese invaded Malaya, all supplies of quinine came under their control, and Krukoff's Guatemalan investment (a struggle in the beginning) suddenly began to come good. His subsequent supply of quinine to the Allied troops in World War Two saved much pain and a great many lives.

Naturally, Krukoff was interested in Iain's work. But he found, as he talked, that Iain had a rather sketchy method of writing up his plant monographs. The traditional monograph not only outlined the plant species, formally classified, but also drew attention to the differences noted in every specimen researched. Iain had come increasingly to feel that this was superfluous. The work this involved for the monograph writer was out of all proportion to the possible gain – and the reader in turn had to plough through overmuch extraneous detail to get to the basic 'keys' of the species. Far more taxonomic research could be done if reference books could be made handier and less detailed, though no less accurate.

But Krukoff hit the herbarium roof! The tried and tested ways of recorded taxonomy were to him the very foundations of botanical science. A monograph should supply everything that a botanist could possibly want to know – and more. It was unthinkable to do it any other way. The disagreement was partly a clash of generations and partly a clash of visions. Iain, who had already used computers, could see that much of the fine detail of cross-referencing and data handling would soon be done by machines. This was, as he had proved in working on his doctoral thesis, a very mechanical business. What the botanist now wanted was an accurate 'handy' reference tool (at five hundred pages not *that* handy) which was authoritative without being encyclopedic. In this way a great deal more new ground could be broken and science carried forward more speedily.

The angry doctor argued strongly and dismissively for the old encyclopedic style. He was used to getting his own way, but Iain stood his ground, putting forward his own arguments. Eventually he was dismissed by his senior with a short grunt and a wave of the hand. The boy had much to learn and Krukoff had no time for such nonsense.

Iain returned to his apartment in New York that evening to tell Anne that he had unwittingly told one of the most powerful men in botany in America that he was out of date in his thinking.

For six months nothing happened and Iain thought no more about it. Then, out of the blue, Krukoff confronted him again. The older man's accent was as thick as ever, but his voice was more subdued. He brought up the subject of monographs.

'I think there is something in what you say,' he conceded. 'The publishers don't like all this detail nowadays. We must look to the future. You are right!'

Iain muttered his thanks for these gracious comments – staggered that this grand, blunt-speaking botanist should even remember who he was, let alone his remarks.

'But that was often the way he worked,' Iain recalls. 'He would think over things for a long, long time. But when he spoke, it was usually worth the wait.'

Krukoff did indeed think many things over while this young British upstart with his voluminous beard (Iain had grown it back, bigger and bolder now) got to grips with the NYBG's collection of Chrysobalanaceae.

He was impressed. In 1967 he asked Iain to accept a B.A. Krukoff Curatorship of Amazonian Botany. The wealthy botanist had decided to endow a post at the New York Botanical Garden and he wanted Iain to hold it. One of his main stipulations was that Iain should spend as much time as possible doing Amazonian research. In fact there would be trouble if he did anything else.

Krukoff referred to the curatorships he endowed – there are several around the world today, including one at Kew – as his 'children for posterity', since he had never had any children of his own. Many thought this an odd, sentimental statement from such a brusque man, but as Iain and Anne got to know him, they came to see the tough exterior was largely a cover for a softer personality which had been much battered by the world events in which he had been caught up. Underneath, Krukoff had a big, old-fashioned Russian soul. Forced to leave home by the communist revolution, he had come to America via Australia and Canada as a lone refugee. Speaking no English, he had started work as a bricklayer and night-watchman. He once told Iain he had had to buy twelve alarm clocks to wake him up each hour of the night for his rounds! He earned enough money to go to Cornell University to study forestry and became a commercial botanist working for the US chemical giant Merck. It was Merck that had set him to work in the Amazon in the thirties to research the chemical constituents of several plants likely to be useful medically.

And so he had become an expert in curare, the notorious blow-gun dart poison used by South American Indians, elements of which are now used as a muscle relaxant in surgery. He was also responsible for isolating the plant derivatives which gave the world atropine. He had learned at first hand what a valuable medical resource Amazonia could be and that there was much of value to learn – especially if the Amerindians were listened to. And this with only minimal research done and relatively few tropical species properly classified. Krukoff encouraged Iain to view the Amazon region in this way, not just as a playground for botanic science but as a valuable – even crucial – world resource, as important as iron ore or mineral deposits, the Scandinavian pine forests or the North American grain harvest. What was not so apparent was the sheer fragility of this resource as a living ecosystem. Like all life it could be killed, and for many species within it there was then no chance of recovery. When this happens, as the saying goes, 'extinction is forever.'

Krukoff had made his home in Guatemala, though he had a second home in Long Island. He invited Iain, Anne and the two girls to stay there briefly *en route* to Brazil for a expeditionary tour. They only made one visit but Krukoff bought a brand-new jeep just for them, and had it delivered to the airport when they arrived, (Krukoff himself could not drive). They stayed at his magnificent hacienda in the hills and met his American wife. The couple had married late in life and so were unable, she told them sadly, to have any children.

Every day Krukoff organized his visitors' holiday picnic schedule as though it were a botanical expedition. And in the evenings he presided over dinner in the hacienda, every bit the Russian-Spanish grandee. He called Iain and Anne 'Don Guillian' and 'Doña Anna'. He especially loved the girls, Rachel and Sarah, and the night before the family left for Manaus he confessed that his greatest wish would have been to have had 'two lovely children like yours, Doña Anna'.

'But then,' he concluded, 'I would only have fought with my children terribly, so it is probably a good thing I do not!' The next day he waved them goodbye with tears rolling down his cheeks. The gruff tyrant of the herbarium had another side which few people ever saw.

'He was,' Iain says, 'a big man in every way – and he had a very big influence on my career.'

9. IN THE FIELD

From 1965 until 1973 – when the Prances spent much longer in Manaus – the longest period of 'jungle time' for the family was one year. This was 1968 to 1969. During this year Iain led four major expeditions into the interior, each one opening up a different facet of the area botanically and enabling him to discover more about the people who lived there. As Krukoff had suggested, the Amerindian peoples knew a great deal about the plants which surrounded them and which they used daily. Already Iain had become fascinated by the variety of cultures he had seen. This year gave him the chance to do a great deal more of this kind of ethnobotanical research.

Iain was also determined to lead collecting and surveying expeditions into the very deepest interior – the areas which were the least collected. Here again he relied heavily on Krukoff's information, often asking about places Krukoff had seen in the 1930s which, though well off the beaten track, might prove a good source of plants and plant data. Not that many tracks had been beaten anywhere, even by the sixties – Iain records that before his own intensive spate of expeditions, 'fewer than ten botanists had ever made extensive collections in the Amazon,' and among these he was tactfully including Dr Howard Irwin, his early mentor, and Dr Bassett Maguire, another old NYBG field campaigner. 'Many areas,' Iain noted in 1968, 'have never been visited by any botanist at all.'

In May of that year the family was re-installed in Manaus. Rachel attended a Portuguese-speaking school. Four projected expeditions were outlined: to Rondonia Territory, in the south of the region; over the border into Bolivia; up the Rio Purus at Lábrea (a tributary of the

Amazon quite close to Manaus); and lastly to Roraima Territory, away to the north, on the border of Venezuela. The expeditions were planned on the information made available to him before leaving New York and later at the Brazilian INPA research institute which had its headquarters in Manaus. INPA's enthusiastic local botanist, Dr William Rodrigues, was, as ever, delighted to see Iain back in Brazil. Each expedition of Iain's seemed almost to double the size of Dr Rodrigues' small herbarium in Manaus, as Iain always remembered to 'put one in for INPA' when collecting his plant numbers.

One innovation which Iain had brought to help this time was an inflatable rubber dinghy. These were just gaining favour with boating enthusiasts in the USA and Iain had quickly seen their potential for river-based exploration: easy to pack and carry, stable, rock- and rapid-proof (they would bounce where dug-out canoes tended to split when they hit rocks in rapids), and above all fast. A rubber boat with a 20hp outboard could plane along at 30mph with five on board; this would greatly reduce the botanists' transit times from camp to research area, speeding up the collecting and pressing process. A canoe could do only a quarter the speed, and carry less weight.

Iain had tried out several different types of inflatable at sea, off Nantucket Island in the USA, running them through fierce tidal rips and thoroughly bumping them about before making his final choice (a Zodiac) for the expeditions. He then took himself to the Johnson outboard motor factory for an engine maintenance course. During the two weeks there, working with the factory experts, he learned all the basics of the motor. It took him right back to his Oxford days and his BSA motorbike. The knowledge he gained subsequently proved its worth many times over, when he was up some obscure jungle creek and the engine of his dug-out or inflatable declined to co-operate.

With his engines, as with his human resources, Iain found that the best way to keep things running smoothly on an expedition was really to get to know them. He is inordinately proud of this modest marine engineer's diploma, and often likes to be introduced at meetings as the only (as far as he knows) professor of botany with a certificate in outboard motor maintenance!

As a prelude to this year of expeditions Iain decided on a two-week test-run up the Rio (river) Urubu to settle the team into the rigours of expedition life and toughen them up for the more arduous trips later. And, as his diary records, it certainly proved to illustrate the common problems and adventures met on all Amazonian field trips, long or short.

Iain is not keen on adventures in the jungle. The expeditions which lack obvious glamour and excitement (which in fact mean trouble) are the best – a smooth operation means that plant collecting is maximized. 'Never let an adventure get in the way of good botany' is his field motto. Mike Balick, one of Iain's colleagues at the New York Botanical Garden, says, 'Expeditions certainly do have glamour and adventure, that's true, but often they add up to long months of solitude, in the rain, away from people you want to be with and care about. All for science and collecting plants.'

Needless to say, the adventures did come, sought or unsought.

For the test run, the expedition party joined the Rio Urubu where the only road out of Manaus, to the nearby city of Itacoatiara, crossed it by ferry – about 120 miles east of the city. On the first day they launched the large transport canoe in which they intended to take the whole party and all the equipment into the interior. It rapidly started to sink. Quickly they hauled it back on land and had it re-caulked where the seams had opened up. Iain had to drive forty miles to Itacoatiara to get the materials, and a day was lost doing the work.

Next day, at the re-launch, it floated with only a modest amount of water leaking in. They strapped on the outboard motor and loaded the canoe with all the expedition equipment and personnel, then paddled mid-stream to start the engine. The canoe swung smartly across the river and continued in a wide arc. Nothing Iain could do with the steering seemed to make any difference. They shut the outboard down again and paddled back to the shore to unpack. The motor was examined. The metal 'skeg' at the base of the propellor was bent out of shape, acting like a rudder jammed to one side. This was the cause of the helpless circling. They waited another day while this too was repaired. Patience, Iain noted on more than one occasion, is always required on expeditions.

Third time lucky, they set off successfully, chugging steadily up-river and negotiating two sets of rapids without problems. Going up rapids is physically hard work, but much safer than coming down. They made camp every night before dark, beaching the canoe and putting up shelters and hammocks, field press and field kitchen. As Iain had intended, this soon became a familiar chore, binding the team together and getting them used to life out of doors.

Each campsite had to be carefully chosen, as the river was in flood and could easily rise a metre or more in the night. They also started collecting and soon got into the daily swing of move, collect, press – a routine which would rule their lives for the next twelve months.

Collecting and pressing is a quite a meticulous business. Plants chosen must be in flower or bearing fruit as these are the prime means of identification. As these rarely occur at the same time collecting may have to be done at different times to ensure a complete specimen, which is why precise notes and locations are important, so the botanist can return to the right place later.

In the rainforest much of the foliage, flowers and fruit are high up in the tree-top canopy, so several tree-climbers are usually employed on expeditions. Gone are the days of chopping trees down. These are fit local people who have perfected the technique of shinning up a tree at remarkable speed with the aid of a sling wrapped tightly round their feet to aid traction. They lop off a branch and bring it down. Some youthful botanists like to do this themselves as tree climbers can made mistakes and cut the wrong tree, but Iain confesses that his climbing days are over now. In any case he maintains that the best use of a botanist's time is actually searching the overgrowth with binoculars ahead of the climbers and spotting the likely trees, lianas (vines) or epiphytes (flowers that grow above ground, usually in the trees). According to former NYBG colleague Brian Boom – now vice-president for Botanical Sciences, Iain had an instinct for this which was almost uncanny.

'We'd be looking for a particular plant or tree and Iain would be walking along and then suddenly stop and raise his binoculars. "I think there may be something there," he would say. There usually was. What you have to realize is that much of what we are looking for, in fact 80 per cent of the rainforest flora, is just green with perhaps small white flowers. At sixty metres or so, even with binoculars, to tell one from another is impressive. Iain usually could.'

When the trees have been spotted, two or three climbers are then sent off to secure the prizes. When they come back with a branch or the whole of a small epiphytic plant the botanist will choose the best parts for identification and put them in newspaper for later pressing. (Iain tells how they often learned a lot of local news that way!) Flowers, leaves and stems can be pressed but usually fruit is too juicy or bulky. This has to be put in formaldehyde instead, to preserve it until it can be dried another way. Back at INPA in Manaus this was done in a large oven.

On the collecting site all the information about each plant 'number' is recorded in the field notebook: where the plant was found, what sort of tree it was on, height, colour, habitat, ecology, and a possible identification. When the specimen reaches the herbarium this collection number is matched with a typed label that will eventually be put on the card with the mounted specimen. The same number goes with each duplicate specimen, wherever its eventual destination.

It may be many years before the specimen is identified, but, once it is, all the old field notes and plant specimens must be connected again to complete the background picture of the find: where it was found, habitat, area distribution, and so on. The number is all-important. Each number may represent a dozen or more similar specimens.

Back in the field the piles of newspapers containing the plants are put between heavy blotting-paper and bound up with aluminium corrugated sheets. These are put in bundles over the kerosene field drier at the main camp and 'cooked' – usually for twelve hours, but more for a very succulent plant. The aluminum corrugate ensures a good circulation of hot air to aid the drying. Once dry, the now dull-brown specimens are packed up into bundles and shipped back to herbaria round the world, with naphthalene moth-balls in the packages to discourage insects from eating the precious specimens.

The work at the field presses is a daily chore, early morning and late evening, for the botanists, and even in the hurly-burly of camp life it has to be done as carefully and precisely as the collecting itself. Mixing up parts of plants, or mismatching fruit or flowers, can cause great confusion when the time comes for identification at the herbarium, and strange and exciting new species have sometimes been 'discovered' following such mix-ups, until closer examination reveals the more mundane truth.

Iain's diary records how wet these first days were, with rain storms most days and the party wearing swimming-trunks much of the time. For some reason, perhaps because the Amazon region straddles the equator, many imagine that it is sunny there, if a little damp underfoot, most of the time. But the rainforest is well named and although there are rainy seasons (March to July in the north, October to January in the south) there is a lot of water a lot of the time. On this occasion Iain records that because of the flooding, the line of the riverbank could only be guessed at by looking at the tree-tops sticking out of the water, like a line of bushy islands. Often, when the river curved sharply, they found themselves steering through their own propellor-wash which had drifted through the trees to re-cross their path.

After four or five days they had settled into field life. Their only doubts related to the navigational ability of their local 'guide', João Padre (John 'Priest', because his bald patch looked like a monk's tonsure). His village had unanimously recommended him as the 'best guide on the river'. This was true up to a point: on the wide parts of the river where they had no real need of him he was fine. It was when the difficult rocky sections came in sight he seemed suddenly full of doubts. Going up had not been too bad. Going down would be more demanding.

At their fourth camp they decided to deploy the inflatable, which duly popped into shape, much to the amazement of Jo,o, who had problems not only with its ability to support him afloat but also with the speed at which it travelled. As they sped off he hung on to the rope lifeline in sheer terror.

In fact, the Zodiac proved its worth even in the brief few days' test they allowed it. But the unhappy João did not. On their return downstream, in the more solid canoe, they hoped his judgment might improve – but it did not. He led them boldly into a section of rapids which he suddenly forgot his way through, halfway down. He later confessed it had been over thirty years since he had come that way. Iain, at the helm, saw his confusion and shouted to the two Brazilian crew at the front of the canoe: did they know a safe route? He couldn't stop, as they were already in the fast-flowing stream and no diversion was possible. The two crewmen responded instantly by pointing across the river – in opposite directions! Iain made a swift decision and drove the boat for a channel he vaguely remembered from the way up. But a submerged log, which he saw only at the last moment, caught the bottom of the canoe and brought it up all-standing, catapulting the two crewmen into the water and breaking open the front of the now helpless craft. Damaged and out of control the canoe sprang sideways towards the bank to bury its broken nose in the bushes. The crew were near enough to swim for it and scramble ashore.

It was a very fortunate escape for everyone. If the canoe had been thrown away from the shore, they would all have been tipped out onto the jagged rocks in the fast-descending stream.

Shakily taking stock, they found that the canoe had been holed and would need repair. But they could not stay where they were so, bailing furiously, they inched their way down the side of the rapids and out into the calmer water below. Even then they needed to travel down-river some way before they could find a landing-place. After half an hour they spotted a possible site, landed, and set up camp to repair the canoe and dry out.

But far from letting the party lick its wounds and relive the experience, Iain immediately called for all those not engaged in boat repair to go out collecting. It was very much his style. He also wisely saw that a couple of hours of interesting field research would be the best antidote for any jangled nerves amongst the botanical party.

At the end of the two weeks the party arrived back at the ferry crossing, already feeling like seasoned campaigners. But one more hazard lay in store, reminding Iain of his own first walk into the jungle. For if the trip had highlighted general boat handling and river navigation as typical Amazon problems, the difficulty of personal navigation in the forest, especially for the scientist absorbed in plant-collecting, was another.

After landing at the ferry, Iain decided that two trips would be needed to carry all the equipment and collected plants back to Manaus, so he drove off with the first load. He returned with Anne, wanting to show her something of the interior. (The girls had been left with friends.) It was supposed to be a gentle introduction to expedition life, to give her an insight into the business of field botany. But when they got back to the ferry an anxious team member told them one of the botanists was reported missing. A storm had swept down on the party which had been out collecting, and one of them could now not be found.

Iain jumped into the inflatable, waving Anne to join him, and they sped off up-river at full speed to search for the missing scientist. They landed and, firing guns in the air and shouting as they went, patrolled the now very sodden trail along which the missing man had been collecting. Eventually, after about an hour, as it was beginning to get dark, they heard a cry. Iain knew three things: that the jungle, like fog, can confuse sound direction; that daylight was disappearing fast; and that the man was isolated in a flooding forest.

Every instinct demanded he plunge immediately into the undergrowth, hacking away with his machete, towards the sound; but instead he coolly went back with Anne to the main landing, twenty minutes' speedboat ride down-river, to collect four more searchers – men from the road maintenance team camped by the ferry. Working as a team, he knew they would have a far better chance of finding the endangered man in the fast-failing light.

And so it proved, but it was a close-run thing. They came upon the missing man, standing dejectedly waist-deep in a patch of flooded forest, in the last glimmer of light. Wisely he had chosen to stay exactly where he was after hearing Iain's shouts answer his own – a difficult thing to do when you are cold and frightened, and the imagination starts suggesting all manner of strange things that might be snaking about in the water around you.

As darkness fell the grateful party relocated their Zodiac (not so easy, since it was black) and steered it – now very overloaded – back down the river to the ferry crossing with the aid of a single pencil flashlight. They then had to cover 120 miles on rough roads in the jeep, back to Manaus before bed. That trial run had enough adventures to last them for the whole year to come.

In the event, the year proved rather less traumatic, and adventures less frequent, with only one more major canoe upset. Here, sadly, Iain lost all the pictures of a new species of *Licania* Chrysobalanaceae which he had discovered, but again no one was hurt and all their equipment was recovered.

All in all that year the four expeditions were a great botanical success. From climbing the high and mysterious Mountains of the Moon (Serra da Lua) in Roraima – which don't appear on any map – and there collecting cloud-drenched mossy epiphytes, to wading the low riverbanks of the Madeira River in Bolivia where floating water-borne plants literally drifted into their arms. They collected 5,700 species, almost 60,000 specimens, not counting wood samples, fungi, ferns and mosses, all sent to herbaria in Brazil, New York and around the world. Iain and his team had cast their collecting net wide across the Amazon basin, and caught much.

Professor Krukoff's new 'child' was certainly turning out to be a true child of the forest.

10. INDIAN COUNTRY

One distinctive feature of the expeditions led by Iain were his research projects into the plant usage of the Amerindians. He had always seen in these tribal peoples an understanding of plant culture. Although it was obviously a direct result of their daily association with the forest, it was also a consequence, he believed, of having to live their lives so 'close' to the earth. And as he had found in Turkey on his first expedition, he felt a personal affinity for such people. They valued plants in a different way from him, but no less highly, and in this they seemed to understand one another.

So it was natural for him to include, even in his earliest work, some observations on the indigenous people's knowledge and use of plants, and note the careful way their resources were managed. Sometimes they simply picked what they needed out of the jungle near their village, though he was quick to observe that the village was usually carefully located in relation to these anticipated needs. But they would also cultivate plants for specific use – a subsistence agriculture – on plots or fields cut out of the forest.

He noticed at first that the Amerindians in these village farms generally avoided cultivating just one crop in a field or plantation area – a monoculture. Instead they mixed all their agricultural needs. Vegetable, fruit and root crops would be grown together with herbs used for medical remedies, or vines grown for the poisons they needed for fishing. Iain noted that this mix of plants would often be very precise.

The reason for this was only partly understood in the wider botanical world when Iain started his researches. Where it was understood, its significance for future exploitation of the jungle areas was overlooked –

with tragic long-term consequences. The unusual nature of the rainforest must be understood if management of it is ever to yield genuine benefits. It will be the only way to keep it alive.

The Amerindians knew that the endless miles of apparently robust, richly stocked, self-perpetuating rainforest is in truth a very fragile ecosystem where everything depends on everything else. Remove one small part and the rest, complex and resilient though it might seem, will collapse too.

Things are different in temperate climates. A flower-bed can easily be made to grow one single variety of flower. Similarly, a large-scale farm in temperate climates can be divided up into fields, each growing, say, corn, or beet, or potatoes. This approach, the monoculture, is both culturally and commercially the most elegant and efficient way of managing the plants in temperate regions. The patchwork of fields and hedgerows in rural England, the tulip fields of Holland, the vineyards of France or the rolling prairies of North America – each has almost become a national symbol.

But monocultivation rests on one thing: the quality of the soil – the amount of nutrients it can carry (though a 'hard' winter which kills insects is also important). Everyone knows about 'poor' soil, soil which has little to offer growing plants. Such areas are usually self-evident. And all farmers can tell you about the areas of poor soil on their holdings – there they can grow nothing.

Many of those who looked into exploiting the rainforest assumed that, judging by the profusion of growth, the soil there must be the richest in the world. Even many botanists (themselves trained mainly in temperate regions) said much the same thing.

But incredibly, the Amazon rainforest has very, very poor soil, despite the fact that in it there is a rich, throbbing habitat, full of exotic species, amazing in its diversity and fecundity. The soil out of which all this grows comes close to that of moondust in its inability to support plant-life without help. But the diversity of species in the rainforest is staggering. In Europe there are no more than half a dozen different tree species in an acre of deciduous woodland. An equivalent area of rainforest would have a *minimum* of 150! It seems that anything and everything grows. Surely that must point to an almost inexhaustible supply of nutrients in the soil?

The nutrient riches of the rainforest are not held in the soil, but in the plants themselves. But not permanently. These nutrients are continuously on the move, up from the forest floor to the top of the trees in the canopy – including the 'air' plants, the epiphytes, the vines, lianas – and down again through dead matter, rotting leaves and fallen fruit. Mycorrhizal fungi then help the living roots of growing trees to absorb the nutrients in the dead matter, and up it goes once more, the cycle starting again, with all in circulation, like blood in the body.

The rainforest is not just a collection of individual plants growing out of the ground near to each other; it is, in an informal non-scientific sense, a total 'organism' in its own right. Even the insects and animals of the forest play a crucial part in the nutrient movements that benefit the plants. And this interaction, this co-operation, goes much further than just passing on nutrients. For in the rainforest the animals, plants, insects and even the fish all form part of a far more complex ecosystem of mutually sustaining life.

For example, some seeds will not germinate unless they have actually passed through the stomach of a certain bird; some fruits will not survive unless protected by a standing army of stinging ants which the plant 'maintains' by excreting nectar for them to eat; some crops can grow successfully only in the vicinity of others which exude a strong scent to repel hungry beetles. Remove bird, ant or scented neighbour and the dependent plant species will die out. Then of course the question is: what was depending on *that*? It does not take long for such an intricate infrastructure to collapse.

Even something as common as the Brazil nut (*Bertholletia excelsa*) in the Lecythidaceae family – another in which Iain has become an expert – has unusual needs. Some recent experiments in germinating a species which Iain had brought back to Kew from Brazil proved a miserable failure, despite the skill and attention of arguably the best gardeners in the world. They had buried them, dropped them, turned them, cut them open, kept them whole, cut the ends off, left the ends on – nothing worked. The nuts stayed in their shells or died, and the director's supply was running out.

Then someone observed that in the rainforest the Brazil nut was usually carried about, and buried, by an agouti. This is a large rodent which, like the squirrel, wisely puts a snack or two by for later. The Kew gardeners made small dents and incisions on the two remaining shells with their pruning-knives, exactly matching the shape of agouti teeth. The nuts germinated! The agouti's larder was clearly indispensable to the plant's survival.

Because of this kind of complex ecology the forest cannot simply be cleared and cultivated in the way that works in the temperate regions – no matter what the sentiment (or commercial pressure) for fields of waving corn in the jungle; no matter how great the need for agricultural produce to relieve social deprivation. The soil alone cannot deliver. So much of the planning – and execution – that went into the 'opening up' of the rainforest of the Amazon basin in the sixties and seventies was based on a fundamental misapprehension. Projects proved unworkable; farms and ranches went bankrupt when government 'start-up' tax incentives ended; and large-scale rice growing, beef cattle rearing (which depended on feedstuffs grown as a monoculture) and even lumber plantations failed, or still need continuing financial support. Not only that, but the cutting and clearing of the forest required to mount these experiments caused permanent devastation. Primary forest can only re-grow out of itself, and even this takes a century of growth.

The Amerindians knew most of this already – as Iain discovered – but no one had bothered to ask them. Later, in the eighties, he found that the few people who had made modest successes of their holdings were Amerindians, or locals who had bought their land from disillusioned farmers, having chosen it carefully, and then planted out a polyculture, relying on the diversity of the forest to produce their reward.

Perhaps the saddest thing about the fate of these Amazon projects, apart from the permanent devastation of the land, is that there were already clear warnings in history. Developers in the 1920s had learned a sharp lesson about monocultures – with rubber. In 1927 the great Henry Ford had wanted an all-American source of rubber for his Model T motor car tyres. He laid out a large plantation in the Tapajóz river region of the Amazon named, with characteristic modesty, Fordlândia. It was a dead loss, despite massive Ford Motor Company investment. Disease leapt easily from tree to tree (instead of being stopped by differing intervening species, the only way rubber trees can grow in the Amazon) and the project collapsed in less than a decade. Later Amazon developers would have done better to ignore Henry Ford's proud words 'history is bunk' and noted instead the harsh facts of the fate of Fordlândia.

When it comes to birthright, the Amerindians have a claim long staked. The lowland Amazon area has been inhabited for about 10,000 years by a population believed to have migrated from the north. Before the conquest of South America by the Europeans, their population is estimated to have been over a million, some say as high as five or six million.

Today it is estimated to be no more than 150,000. In Amazonia the survivors have been driven back further and further into the hinterland by each wave of colonists. In 1900 there were about 230 tribal groups recorded, but during the early decades of this century many tribes were decimated by the exploitation of the rubber industry. Today there are fewer than 140 tribal groups, which indicates that over ninety groups – with all their manners, customs and unique understanding of the forest – have perished during this century. The survivors still retain their tribal groups, but many consist of fewer than a hundred people, too few to be a viable stable unit. These too are on the verge of extinction.

But the Amerindians are a remarkable people who have integrated themselves thoroughly into the Amazon ecosystem. Some tribes depend mainly on agriculture for their existence, others are nomadic. Some depend largely on the forest, others upon the river and fishing. There are about thirty known tribes that remain completely isolated from contact with Westerners. When this contact does occur it seems to affect some tribes more than others – hence the pitiful sight of some becoming beggars for the throw-aways of the city dwellers, whereas others stick with the lifestyle they have always had, taking from western culture what suits them and leaving what does not. The Amerindians remain distinct from the Brazilian immigrants to the Amazon region, who are known as the 'Caboclo' people. They are the local 'Amazonian Brazilian', usually a mix of European, African and Amerindian blood. In the main, the Caboclos are riverbank dwellers (the rivers are the main highways of the region, even today) and are often freelance rubber tappers or nut gatherers, or simply local traders. Portuguese-speaking, they are a very hospitable people who live on very little, raising their families in small huts on the river shore. Sometimes they even build these huts on rafts off-shore, connected to the land with a gangplank so that their homes can rise and fall on the flood water. They trade by taking local produce to market by canoe, often many miles up- and down-river. It is a subsistence living and their diet is based on the cassava plant supplemented by rice and beans and the occasional egg from a kept chicken, available fruit, such as bananas, and, of course, a plentiful supply of fish.

These are the people whose puttering motors announced the dawn for Anne in the city of Manaus, as they made their way to market, canoes piled high with the wares they hoped to sell. They also form a great part of the population of the city itself.

In his first ten years of expeditions Iain visited more than a dozen Amerindian tribes as part of his research work: from the Mayongong and Sanama Yanomami in Auaris, Roraima Territory (now a State), and the Makú of the upper Rio Negro (both in the north of the region) to the Karitiana and Paacas Novas in Rondonia in the more frequented south.

He soon came to value their help personally as tree-climbers and jungle guides, as their knowledge of paths and trails proved invaluable in speeding up collecting. They were usually delighted to show off their intimate knowledge of the flora and fauna of the jungle, though some feared that the 'magic' of their plants would leave if information was revealed. This was, for the Amerindians, hard-won knowledge and tribal lore which, Iain was determined, should be fully credited to them in cash or kind whenever western science or commerce made use of it. This soon became a standard rule for all the ethnobotanical research he oversaw at New York and still is so today at Kew. This is clearly the Amerindian's intellectual property.

Apart from the people themselves, Iain always mentions the strong impression that the communal tribal buildings, or malocas, made on him: 'dark, lofty and imposing'. These are basically very large huts, made from rough-hewn poles and thatch from the forest – a single hut housing the whole village.

The malocas may be round, oval or rectangular, completely roofed or open in the centre. The high top of the maloca was usually home to rats, cockroaches and other insects that liked to hide in the thatch, even occasionally a snake or two, though these were usually chased out pretty quickly or killed by the occupants.

The Mayongong maloca is round or oblong, about ten metres wide by forty long. The walls are made of mud, packed into a framework of small twigs, wattle and daub style. The roof is thatched with large palm leaves and there are several small doors around the outside of the circular wall. Each door leads from the outside directly into a small room or 'segment' for one family unit. This is divided by thatch walls from the one next door. Another door leads off inwards to the large circular central meeting-place, which is used by the whole community and for visitors.

The family in a segment-room is a man, woman and their children plus any members of their extended family. Each room has a fire which is the focal point for family life. It is used for cooking during the day and for warmth during the night. Each family member slings a hammock round the open hearth and a supply of wood is left by the fire so that it can be kept burning during the night. When visiting the Mayangong Iain and his team were provided with a pile of fuel for their fire *and* a boy to keep it going for them throughout the night.

The Amerindians go to considerable trouble to keep their fires alight because of the great difficulty of re-lighting them in the permanently damp rainforest. On the occasions when it's necessary, they make fire in the standard boy scout manner by spinning one stick against another in dry tinder. But it takes a long time. A smouldering tree trunk in a clearing which has been burned over for crop planting (a procedure known as slash and burn) is a valuable source of fire for the tribe for many days. Fanning it with a palm leaf soon brings it back to life. Smouldering wood was also often carried from one place to another, and woe betide the traveller who let a hot brand go out *en route*. This was viewed as serious negligence and a breach of tribal etiquette.

The many domestic fires inside one building caused a great deal of smoke and Iain remembers the maloca atmosphere, day or night, as always very smoky and dark, producing stinging eyes and coughing fits among visitors. There was usually very little ventilation apart from the exterior doors, though in some there was a trapdoor in the roof, pulled open by a vine rope run over a high beam, and worked from the floor. But at night this and the outside doors were always closed for protection from wandering animals and from other tribes, and the 'fug' really built up.

'Then,' says Iain, 'it was absolutely stifling, and the shapes of people and animals could only dimly be seen moving about by the red, smoky light of the flickering fires.'

He contrasts this with the open-centred type of maloca favoured by some of the Yanomami. This is essentially a circle of roofed dwellings open in the centre. Each family adopts a section (there are no thatch dividing walls) and again has a fire as a focus. Their possessions hang round the outside wall which has some exterior doors. The open area in the middle is used for ceremonial dances and rituals, and as a children's play area. Iain says that nothing ever grew there, it was just sun-baked dirt, the village fields being nearby outside the circle of the maloca dwelling.

There is a complete lack of privacy. Anyone who stands in the centre is visible to everyone around the side and equally everyone around the circle is exposed from the centre. Iain quite enjoyed this as his prime business was to watch and note what went on.

'The daily routine of the village runs before your eyes. The women can be seen preparing the daily stew, while the men who are not away hunting are resting in their hammocks or fighting. The shaman is dancing or making his visits to the sick, or one Amerindian is applying body-paint to another. The only items of furniture they have are their vine hammocks and a few primitive shelves on which they keep cooking utensils and sometimes a bunch of bananas or cassava roots.'The air was fresher and cleaner in the open type of dwelling and the light better, but there were still drawbacks.

'The Yanomami often keep dogs as pets for hunting. The result is that the floors are thick with fleas. After only a few minutes inside the maloca you can be literally covered with fleas.'

To Iain, slinging his hammock in a smoky maloca or swatting at dog fleas whilst he watched tribal life go on round him was all part of life's rich pageant on jungle expeditions. Other botanists are not quite so fond of this ethnic element of their trade and prefer to stick to plants. But Iain is notorious for never being able to resist 'just one more trip' to see more of the forest people whose company he so much enjoys.

11. FOREST FARE

The Amerindian tribes that depend on shifting agriculture invariably use the 'slash and burn' technique: cutting down enough forest for a field and then setting fire to the wood. The crops are then planted in the new, fire-blackened clearing. The felled area is the farm limit and the wood ash provides the all important nutrients for the future crops, as the soil cannot.

A typical slash-and-burn field of this type ranges from 2.5–15 acres, varying with the size of village or family it must support. The main crops are usually the staple cassava or manioc – *Manihot esculenta* – and bananas, but much else is mixed in besides. A forest farm like this usually yields crops for a maximum of four years. After this the land is exhausted and they must select a new area and abandon the old. The original primary rainforest will take about a hundred years to regenerate on the cleared area, though a dense secondary forest will spring up within about five.

The Amerindians cut these fields from within virgin forest, so seeds will move in quickly from the adjacent primary jungle when the land is left. The areas cut are small, as are the Amerindian numbers, and secondary forest is seldom cut down, so the local ecology is fully sustained. Recent carbon analysis has shown that surprisingly large parts of the rainforest have been cut, burned, farmed and then permitted to regenerate over the centuries. Very little is wholly untouched virgin forest, though it has been so well-managed by the Amerindian peoples over the years that most of the recovered primary forest is undetectable from forest which has never been touched.

Western-style forest clearance is very different. Large swathes (many hundreds of thousands of acres) of land are bulldozed for the road projects, mining projects, lumber projects and associated agricultural settlements which, when they fail, offer no chance of regeneration. What is left turns into a desert scrubland, devoid of beauty and value. Much botanical expertise is being directed today to try to discover ways of reforesting these man-made deserts.

The jungle is cold store, supermarket, tool shop, armoury, boat chandlers and pharmacy all in one to the Amerindians. When on a hike they rarely take a haversack of food with them. They know the trees and use them as guides to fruit, berries and nuts, or will watch for plants which form the diet of animals they want to hunt, then lie in wait and catch them. They rarely go hungry, and travel fast and light.

Iain talks about his visit to the Jamamadis on the Purus River, some 200 miles from Manaus as an example of the 'total plant culture' of some Amerindians. Research has shown that most Amazonian tribes name and use some 75–95 per cent of the trees and plants in their area and often have more than one use for each. Over a three-week period Iain noted in detail this kind of comprehensive plant use by the Jamamadi.

Their individual houses (they did not use malocas) were constructed on a wood frame with palms placed on top as thatch, the walls and floors being made from durable but flexible timber, the frame bound together with vines and stringy fibres from the forest. The roof palm, known by the Jamamadi as 'Ubim' was of the genus *Geonoma*; the floors and walls were made from the trunk of the 'Paxiuba' palm, *Socratea exorrhiza*. This palm was particularly useful because the wood could be split easily into long, tough strips.

Their canoes were dug-outs made from one of three species of tree, two of them members of the laurel family: 'Itauba' *Mezilaurus itauba* and 'Lauromiri' *Aniba santalodora*. All dug-out designs avoid the need for any securing or caulking materials so for a watertight hull this wood has 'grown-in' integrity and strength. Their paddles were also made from 'Itauba' wood or from the so-called 'paddle wood tree', *Aspidosperma*. This is used by many tribes and a mature tree grows with exterior 'buttresses' which fan out from the main trunk like fins on a rocket. One tree can provide several strong, flat pre-shaped paddles, by cutting down along the line of the trunk and then out across the base of the buttress.

For baskets, panniers, trays and sieves the Jamamadi used a stringy vine they called 'Titica', *Heteropsis jenmanni*, which belongs to the Araceae family. Iain notes that this also is something used throughout the Amazon. It is first split up and dried into 'strings', then loosely woven into baskets. For stronger rope the Jamamadi used the bark of several members of the *Annona* family, stripping it off in fibrous lengths, like webbing, to make up items such as baby-carrying slings or ties for their house frames.

They have an interesting use too for one of Iain's favourites, the species of Chrysobalanaceae – *Licania octandra* – known by the Jamamadis as the 'Caripe' tree. It is very rich in silica. They strip the bark off the tree and make a fire of it, ensuring no other wood is used. The ash is left to cool, sieved, and the resulting powder mixed with clay. The pottery made out of this, when fired, comes out very much harder and stronger than ordinary clay pottery and lasts many times longer. It is the primitive equivalent of 'Pyrex' glassware and is heat resistant so it can be used on the fire for cooking. This pottery can always be told by the metallic ring it gives when flicked with a finger.

The Amerindians' most notorious use of plants is in the manufacture of 'curare' poison, used to tip the darts of their blow-guns. They also used plant-derived poison to catch fish. Iain studied both.

Curare for the blow-gun, he found, tended to have a different recipe in each tribe, though a common base is extracted from the vine *Strychnos* which the Jamamadi called 'ira'. In their case this was mixed with three others: 'bicava' *Curarea toxicofera*, 'boa' *Guatteria megalophylla* and 'balala' *Fagara*. The tips of their blow-gun darts are dipped in this mixture and, when fired, the impact of the dart delivers it into the bloodstream of the hunted animal, paralyzing the muscles. The poison rarely kills, unless the animal is very small, but once immobilized, the hunters' quarry is then easily dispatched in a more conventional way.

The poison is safe for the consumer (a matter of some concern to Iain as the 'honoured tribal guest') as cooking degrades it. It is also effective only when delivered straight into the bloodstream and not when passed through an eater's digestive tract. Iain did not suffer any unpleasant consequences when he was asked to join the feast, and Fred Orr – the Christian missionary who worked with the Jamamadis and helped Iain throughout his stay (particularly with translation) – said he had never known anyone suffer from eating animals which had been brought down with a poisoned blow-gun dart.

The blow-gun itself is made from the 'armoury' of the forest which also supplies bows, arrows and spears. The Jamamadi make their blow-guns out of the nutmeg family *Iryanthera tricornis*. When the tree is cut (it has to be young, or it will be too large as a weapon) the trunk is slit down the middle, into two halves, and carefully hollowed out very smoothly to make the 'barrel'. The halves are then bound back together with string-like fibres from the inner bark of the annonaceous tree *Envira*, and glued with resin from another, *Protium*. The whole gun is some three metres long, with a remarkably smooth bore about 4cm in diameter. Iain estimates it is accurate to a range of around forty metres, depending on the skill of the huntsman, and, he notes, 'it is deadly silent'.

The darts are made from sharpened half-metre lengths of the 'paxiuba' palm, *Socratea exorrhiza*, the same as is used for hut floors and walls. These are then bound with twisted raw cotton which provides the compression in the pipe and also some stability in flight. The darts are always poisoned before use.

As for bows and arrows, the Yanomami, another tribe Iain has studied in detail, have made an unconscious revision of the old English longbow. Theirs is the long-arrow! This is made out of the stiff grass *Gynerium sagittatum* usually cultivated specially for the purpose. At two and half metres, it is longer than the bow which launches it and much taller than the warrior who fires it. But its great length makes it highly accurate, and the Yanomami generally prefer this to the blow-gun for hunting. A groove is cut into the tip of the longarrow to increase the capacity of the curare it holds.

Game is hunted by individuals or warrior groups but the poisoning of fish is usually a formal or even a ritual affair involving the whole village. Iain believes that this is part of the very strong 'conservation consciousness' of Amerindians. They know that the fish stocks must be renewable or the tribe will perish, so they limit poisoning to a ceremonial occasion once or twice a year, which ensures that stocks recover.

The Jamamadi take their fish poison from a member of the bean family, a vine they call 'cuna', *Derris latifolia*. This is cultivated in the village field and is extremely easy to propagate, as cuttings taken from the root sprout very quickly. When a fish harvest is deemed appropriate by the village shaman, the cuna is harvested, root and stem, cut up into short sections and then mashed by a heavy club. The broken pieces are lowered into the chosen stream and the water stirred up to wash out the juices. As the tainted water flows downstream the poisoned fish float to the surface where they are collected further down in baskets.

Almost everyone helps in the collecting, including children, who love to splash around in the water and regard the whole thing as an exciting and eventful day out.

The poison drifts downstream for about half a mile before becoming too dilute to take effect. Iain noted that fish only partially affected usually seemed to recover. Fred Orr told him that he had never seen fish caught in this way poison any of the tribal people. Even those splashing in the water were unaffected. Iain discovered later that this was because the fish die from paralysis of the gill system: the poison does not enter the bloodstream.

Every scrap of food collected is eaten, even down to the smallest fish. The whole tribe eat huge quantities to 'last' until the next time, often many months later. In the interim, they hunt and feed off the forest in the usual way. Iain was told that the fish streams were used in strict rotation, to ensure the regeneration of stocks. On one occasion, with the Makú people, he marched for eight hours with one village group, carrying the prepared poison, crossing several streams until the 'right' one was found. Living close to the forest and its resources means that the Amerindians take conservation very seriously indeed.

'We (in the industrialized countries) depend on plant and animal resources every bit as much,' Iain writes. 'It may not seem so, but we are living off our capital as regards the environment and that is as unsustainable in the environment as it is in our bank accounts.'

It is a sobering analogy, for the fish and forest stocks are the bank account of the tribal people, and they know with mortal certainty that they must take great pains to keep it in credit.

Finally, to go with the fish and meat, the Jamamadi grow crops in their fields, including several types of banana, sweet and bitter cassava, pineapples, yams, taros, corn, cashew nuts, papayas, lemons and mangoes.

Medicines too come from the same two sources, forest and field. Iain noted plant cures for coughs, sore throats, rheumatism, worms, and even toothache. He tried this last, chewing the root offered him and finding there was a distinct tingling and numbing effect in his mouth; evidently a mild anaesthetic. A firm advocate of plant research in the interests of medicine (he himself has benefited from many Amerindian remedies over the years), Iain nonetheless cautions against too quick an acceptance of the cures. On one occasion when he was noting the medical contents of one tribe's plant 'dispensary' he was shown a leaf which the shaman assured him was excellent for curing aching backs. Iain carefully wrote down the species. Fortunately he then asked how the remedy was applied. To illustrate, the shaman picked up the leaves and waved them over the shoulders of a nearby sufferer while pronouncing several loud ritual injunctions. He then declared the pain banished. Iain decided not to pursue research into the active ingredient of that particular plant cure!

But there was no doubting the immediate efficacy of the compound the Jamamadi used as 'snuff'. It clearly packed more of a punch than the elegant variety sniffed in the drawing-rooms of nineteenth-century Europe, though it had similar origins: the tobacco plant *Nicotiana tabacum*. This the Jamamadi dry and mix with ash from the bark of the 'cupui' tree, *Theobroma subincanum* in roughly equal quantities, then pulverize. Called 'shina', the powdery result is sniffed up each nostril with the aid of a snuff tube, usually a hollowed-out monkey bone. Iain observed that the recipient very soon began to get light-headed and later became quite intoxicated.

Most Jamamadis keep their own small 'shina' supply handy, though it is generally used only in the evenings or after work, a sort of nasal equivalent of 'a drink before supper'. To Iain the most concerning thing was the number of children who had acquired the habit early in their lives. He records one little girl of four regularly taking sniffs from her own 'shina' bottle.

He recorded a more extraordinary use of hallucinogens during a trip to the Sanama at Auaris. They used the virola tree – *Virola theiodora*, a common source of hallucinogens in the Amazon – at funerals. Iain was present at one of these, for a warrior who had died of the common cold, something to which the Amerindians have little or no resistance.

The drug is prepared by heating up strips of virola bark until the resin oozes out. This is then scraped onto arrow heads and left to dry. These arrow heads are not used in hunting but solely for ceremonial storage of the drug. On the appropriate day (about half-way through the funeral ceremony, which can last for up to two weeks) the snuff is made ready by scraping some of the resin off the arrow heads. This is then sniffed up through a small pipe, one tribesman often blowing the snuff up another's nostril.

When the hallucinogen takes effect the Amerindians begin to dance in the centre of the maloca, waving their weapons above their heads. Then comes the strangest part of the ceremony, at least to our eyes. Anyone bearing a grudge against another, or who feels offended in any way, offers his chest up as a target. This is hit, often with a fist, but possibly also with a stone or other sharp item, one blow only, by another tribesman. The blow is then returned. This continues, with blow traded for blow, until one of the two gives up. Although blood often flows, the opponents do not appear to flinch or feel much pain, as the drug seems to act as an anaesthetic. But that is not all. They then engage in a mutual screaming match.

Squatting down to face each other, they grasp each other's shoulders and, leaning forward, bellow as loudly as they can into each other's ears. At one of the ceremonies, Iain records that the noise was so loud 'that it was difficult to remain in the maloca'. Once again the hallucinogen seemed to deaden the effect of this mutual violence. At the high point of the ceremony, when the noise was loudest, the ashes of the dead man were poured onto a central fire in the maloca. Then, as the drug wore off, the noise gradually diminished. 'A very memorable experience,' wrote Iain afterwards.

Plant-based body paint is often associated with such ceremonies and tribal life in general. Iain has never visited a tribe that did not use it in one form or another. The most common is a red dye made from the urucum or achiote plant, *Bixa orellana*, the colour coming from the bright red aril that surrounds the seeds. The plant is also known as the spice annatto, increasingly used in the west as a less chemically toxic 'reddener' than some manufactured colourings. This is used by warriors in conjunction with bead necklaces of seeds and head-dresses of leaves, wood, bark and feathers.

A blue or blue-black dye which the Yanomami use as a form of tattoo (though indelible, its subcutaneous effect lasts for a few weeks only, not for life) is the genipapo tree, *Genipa americana*. In contrast to the warrior's paint, this is often lovingly and painstakingly applied by a mother to the face or body of her child. Its intricate pattern then becomes a personal statement of their relationship within the general community of the tribe. Iain observed that there was never any interest among the Amerindians for green or yellow dyes, though there are certainly plants to provide them. Red and blue, and occasionally white or black, are the favourites.

'But then,' he comments, 'if I lived in a jungle full of green foliage and yellow flowers I would probably be looking for some other colours too!'

As well as visiting villages to make ethnobotanical observations, Iain worked with Amerindians to further his direct botanical quests. Many of the tribespeople were willing to join him in collecting in return for such unusual treats as a ride at thirty knots in the rubber Zodiac. One tribal chief liked this so much he asked if Iain could take his family out for a trip too.

Iain realized that this actually meant his whole tribe and so had to set aside the rest of the day for it, with one of the student members of his expedition taking out groups of gleeful Amerindians five at a time to do figures of eight, up- and down-river. But he was happy to do it, knowing that for these tribespeople this was the trip of a lifetime and many of them would spend considerable time with him in return, showing him good areas for plants, climbing trees, or finding flowers, herbs or fungi he particularly wanted to collect. Iain was only too pleased to trade a day 'messing about in boats' in return for this kind of help.

However, one particular collecting expedition in 1971 worked out with this kind of trade in mind, proved a considerable challenge both for the Amerindians involved and for Iain himself.

Iain's idea was to trek for 170 miles between two airstrips in Roraima Territory, by the northern Brazilian border with Venezuela. It would, he estimated, take about three weeks and provide unique collecting opportunity in the region.

He and Bill Steward, an American undergraduate student, and three Brazilian field helpers, would fly into a small mission airstrip at Serra dos Surucucus. There they would meet the local missionary, Fritz Herter, and walk the 170 miles to Waiká airfield, with a group of Amerindians as helpers and porters (for the collected plant specimens). Then they would be treated to a plane ride back to their village near Surucucus. Basic food and supplies would have to be carried with them, to be supplemented by game hunted by the Amerindians who were skilled in living off the forest.

The scheme looked good to Iain, so the planning was swiftly completed and the team flew out to Surucucus. Nineteen local Yanomami tribesmen were recruited and Fritz, who had worked many years with the local people, joined the march as interpreter and guide.

All progressed well enough at the start as they strode out along the trail, establishing a routine of half a day walking and half a day plant collecting. Iain was particularly pleased not only to be back in the jungle with the Amerindians (they stopped along the way in Yanomami villages and Iain made many notes and collections there), but also to have the company of Fritz, a practical Christian man who enjoyed and respected the forest and forest people as much as he did.

94

There was just one small matter of concern: one of the Brazilian field men, they discovered, did not share their view of Amerindians. Worse, he made no secret of it, making pointed exits from their camp when they stopped on the trail and walking out of the maloca making rude comments when the party stayed overnight in villages.

Since everyone else was getting on well, this behaviour stood out. It was not long before the Amerindians took the offence intended. A week into the walk, they decided they would have to kill him.

The first Iain knew about this was being woken suddenly by an almighty racket in the campsite at a very dark hour of the early morning, when everyone ought to have been asleep. It was the loud calling of Amerindian death chants and the simultaneous slapping of machetes (flat part only, to start with) on the bare skin of the terrified field worker who had been summarily turned out of his hammock, surrounded by hostile tribesmen, and faced with imminent execution.

The situation was little short of desperate. Iain knew enough about the Yanomami to know that if they had decided on something, by the rules of the tribe there was now no going back. If things were going to turn violent he, as leader, had a clear duty to try and stop it. But if he tried to intervene physically, who knew where it might end?

At that moment Fritz, who had also been startled awake by the racket, strode unhesitatingly into the midst of the mêlée and began to harangue the tribesmen in a high-toned language that Iain did not recognize. After a few moments the chanting stopped and they began to listen. Then they began to dispute with Fritz in the same obscure sing-song tones. After some time the still shaking Brazilian was reluctantly released and the Amerindians sloped off quietly to their hammocks.

'What on earth's going on Fritz?' asked a thoroughly bemused Iain when the missionary at last came over to speak to him.

'They were working up to killing the fellow,' replied Fritz shortly. 'I've seen it before. Once they decide, there isn't usually much you can do. But I challenged their leader in the language of the High Council. It was the only thing I could think of. It's a kind of diplomatic language they use for serious tribal problems, life and death cases, that kind of thing. Thank the Lord they stopped to listen.'

Fritz then explained that he had told the Amerindians that the Brazilian, though apparently rude and hostile, was in fact an ignorant and uncivilized friend of Iain's who knew no better, and it would be less than courtesy to kill him, whatever they thought of him personally. In the end the Yanomami had taken the point, and not wishing the tribe to be viewed in a bad light had called off the attack. Iain was suitably grateful to Fritz for his quick thinking and accurate assessment of the situation. 'Murder On The Amazon' was not the kind of title he wanted to head his field report to the New York Botanical Garden.

At the end of three weeks the two dozen walkers staggered out onto the airstrip at Waiká burdened under the weight of the many hundreds of pounds of plant matter they carried in their wickerwork forest-wood backpacks. The collecting had gone very well. The wrapped specimens were all thoroughly soaked in formaldehyde to preserve them until they could be flown back and pressed in Manaus and one proud Amerindian carried atop his pack the incongruous white plastic baby bath which had been the essential instrument for this preservation.

The aircraft met them two days later and, in relays, flew the nineteen excited Amerindians, and Fritz, back to Surucucus. By all reports they made the most of it, craning out of the windows of the small plane as it swooped low over their jungle home for all the world like rubber-necking tourists. Trading plant collecting for a plane ride had been a very good deal.

When they had all gone Iain decided that, on the food still available, the remaining team of five could manage to spend a few days at Waiká doing a little more collecting. This they did, finishing the food and the collecting four days later. Eagerly they awaited the promised last plane home.

None came. Three days later, having culled some food from the forest, they heard the drone of twin engines over the canopy and spotted a plane approaching to land. They saw right away that it was not their usual one, but ran towards the airfield clearing nonetheless. As they emerged from the forest they were shocked to see the aircraft crumple down onto the runway and skid towards them, propellors kicking up dirt and mud as it shuddered to a halt.

Three very pale passengers and a pilot got out gingerly. Fortunately all were unharmed. One was the botanists' regular pilot, Lyn Entz. Iain asked him what had happened. Lyn explained that the insurance cover on his plane had expired and it had therefore been grounded until new cover could be agreed. He had come out on this one as navigator. He wished he had been flying it, he added quietly, he might have saved on someone else's insurance!

A Yanomami Indian youth at Auaris, Roraima, Brazil

This Yanomami wife is decorating her husband with body paint made from a species of *Pourouma* before he takes part in a ceremony

A group of Yanomami Indians on the trail from Surucucus to Waiká

A Yanomami Indian village house, or maloca

A Mayongong Indian house being built. They use timber of many different species of trees for the component parts of their house, termed a maloca

A Paumari Indian stripping the basketry fibre from a plant of *Ischnosiphon*

A Mayongong Indian making a basket from the fibre of the aerial roots of the aroid *Heteropsis jenmani*

A Bora Indian woman painting Sarah Prance with body paint made from the fruit of the genipap (*Genipa americana*). This dye does not wash off and remained for two weeks!

The eye-like appearance of the fruit of guaraná gave rise to the Maués Indian legend that its origin was from the planting of the eye of a child warrior and hero

The Chrysobalanaceae is a family of large shrubs and trees growing in the tropics. Iain made it the focus of his research. This is a tree of the genus *Parinari*

The flower of *Couepia chrysocalyx* in the Chrysobalanaceae which is pollinated by moths

A Yanomami Indian in a canoe on the Rio Uraricoera near Waiká

The Yanomami Indian way of starting a fire. Re-lighting fires is very difficult in the permanently damp rainforest

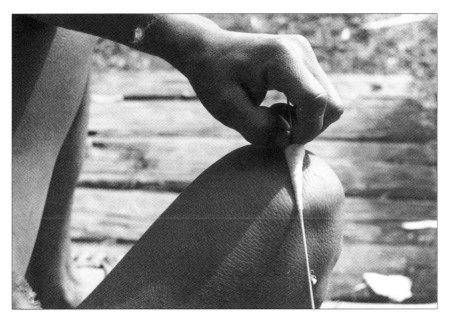

Preparing the wad of a blow-gun dart. The darts are tipped with 'curare' – a plant-derived poison – which is also used as a muscle relaxant by surgeons

A Jagua Indian with his blow-gun and dart. This tribe wears grass skirts and similar dressing may have led to the accounts of tribes of women warriors or Amazons that gave their name to the river

Iain and a *Victoria* water lily

An Amazon water lily (*Victoria amazonica*) flower under investigation – note the thermometer

Iain in the rainforests of
Brunei with John
Dransfield examining a
rattan palm

Yanomami Indians at Toototobi taking part in the cermonial dance to welcome
visitors

A Maku Indian fish-poisoning ceremony. The poisons are plant-derived, in this case from the spurge *Euphorbia cotinifolia*

A Sanama Yanomami boy with cassava bread. Cassava plant has starchy roots and is the staple diet of most Indian tribes and of the local residents of Amazonia

A rubber tapper coagulates the rubber by pouring on the latex and spinning a ball of rubber in acid smoke

Sapucaia nuts (*Lecythis pisonis*)

A Brazil nut gatherer opening the outer case with his machete

'Slash and burn' – a patch of forest that has been cut down for the planting of crops. This will do well for one or two years and then the soil will be exhausted

An agrovila in 1973 along the Transamazon Highway. This demonstration village was also a demonstration of how not to manage Amazonian soils

Taking the expedition canoe up a rapid on the Xingú river, Pará, Brazil

Hauling one of the inflatable boats up a rapid on the Curuquetê river is hard and dangerous work

Iain took an outboard mechanics course to learn how to keep the expedition motors running

A Dení woman preparing their contraceptive from a vine of the genus *Curarea*

The Brazil nut flower (*Bertholletia excelsa*)

(opposite top) A Paumari Indian stripping the bark of *Tanaecium nocturnum* to get hallucinogenic drugs

(opposite bottom) A shouting ceremony of the Yanomami under influence of their hallucinogenic snuff

This young fisherman had just caught a piraracu (*Arapaima gigas*). This is a small example of the Amazon's largest fish

Iain soon discovered the crashed plane was piloted by an ex-Brazilian Airforce colonel who had been taking an aerial survey team out to map the forest. He had been asked if he would bring the botanists out as, with Lyn grounded, they would be getting short of food. He had generously agreed to do this but evidently was not very experienced on jungle airstrips and his plane was now damaged beyond repair. Iain had to decide quickly what to do, as the whole party was clearly marooned in the jungle. As expedition leader he was responsible.

He thought that there was a reasonable hope of air rescue. So he set everyone to pushing, pulling, and dragging the damaged plane clear of the runway. There would certainly be no rescue if no one could land. Then food became top priority and Iain organized them into two teams, foragers and fishers. Taking a leaf out of the Amerindians' book, he set them to scour the 'larder' of the forest. Unfortunately they had by then all safely returned home, so it was down to Iain's own botanical know-how to choose from this larder correctly. The foragers he led himself, having decided to collect fruit from three of the trees in the area. Most significant was Pupunha palm, *Guilielma speciosa*, which has small, fleshy, large-seeded fruit. When boiled for no less than four hours they soften into a chestnut-flavoured mouthful. These were supplemented with fruit of *Belucia* (a shrub of the Melastomataceae family), palm hearts, and what the fishing party brought in. They soon reported back with several dozen piranha! As emergency diets go, it was not at all bad, with the piranha proving rather bony but quite good to taste and the punpunha wholesome if rather monotonous.

Iain recalls that, for the most part, the survivors were calm and disciplined about the stranding – or if they were worried, they did not show it. Most jumpy were the surveyors, who had never been out in the jungle before, let alone cast away in it. Their nervousness increased by the day and Iain began to wonder if he would soon have to take strong measures to calm them down. Fortunately, on the third day the sound of an engine was heard and shortly after a light plane hopped down to land bringing a nurse and mechanic to the rescue.

As no one was injured, and the other plane was beyond repair, though welcome, these two were not of immediate help. So Lyn Entz climbed aboard and, insisting he flew the plane, took off, leaving the nurse behind and promising to be back with his larger machine at the very earliest opportunity.

It took him five days to sort out the red tape and fly back. By this time the stranded surveyors, botanists, nurse and airforce pilot could barely look at Pupunha palm fruit. In fact, after eating about three dozen of them three meals a day for nine days, none of them could face eating it at all for over a year afterwards! But it had kept them alive and, as at last they flew out of Waiká airstrip with Lyn at the controls, they all felt that even with the modest skills at their disposal the forest had after all supported them in their extremity, as it has so many who have depended upon it down the years.

12. WATERSHED

The long hike on the Waiká trail and the unfortunate loss of the aircraft had certainly tested the resourcefulness of Iain's team more than they had anticipated. Despite this, they made an excellent series of collections, breaking new ground. Even the life and death drama of the fieldworker had ended safely, if not exactly happily, thanks to the experience and wisdom of missionary Fritz Herter. (The Brazilian himself of course felt that all his opinions of Amerindians had been doubly confirmed.) All in all a journey that might so easily have ended in disaster had been turned into a great success.

But within three months Iain was to set out on another expedition and this time there was a far more serious challenge to face. Matters would get so serious that, for the only time in Iain's field career, an expedition for which he was responsible would come to a dead stop. It would be all they could do to survive.

'It was a time,' Iain says in retrospect, 'when I truly believe but for the grace of God and the presence of one or two of his servants – in the shape of local missionaries – I, and possibly a number of my companions, would not be here to tell the tale.'

Serious comment for a man as experienced as Iain in facing and solving problems of field work in the Amazon.

It was to be the third expedition planned for 1971 – a 2,400 mile round trip by boat and boot up the Rio Purus and down the Rio Madeira, midway crossing over the watershed between the two rivers. This area, the south-west rim of the Amazon basin, is the Brazilian equivalent of darkest Africa.

In the centre of the Amazon region lies Manaus. It is a large conurbation and offers all the support services of western civilization: air and sea transport, radio and telephone communications, hospitals, government institutions and so on. This means that in the geographical heart of the forest (which one might imagine as remote) there is a very accessible, modern city. It is only when one goes out to the 'rim' of the region, off the beaten track (which Iain, of course, was intent on doing), that things become more remote. The south-west rim of the Amazon basin – the middle of the whole South American continent – is arguably the most remote area of them all.

The Rio Purus and the Rio Madeira rivers run for some 1,500 miles north east from their sources in the highlands of Peru and Bolivia in the west, slanting up to the east-west line of the Amazon some 200 miles apart and reaching the Amazon river about 100 miles either side of Manaus. Iain's intention was to travel up the Rio Purus and then the Rio Ituxi, which comes directly from it, stopping frequently to collect, and then take a smaller river, the Rio Curuquetê, upstream towards the watershed. Then they would have to hike across this, and, once again resorting to boats, ride the Rio Madeira back down to the Amazon and home.

This whole course would roughly resemble a displaced oblong with 200 miles of the Amazon at the top (with Manaus in the middle), 200 miles of minor river and forest trail at the bottom (with the watershed in the middle), and 1,000 miles of river passage, out and back, either side.

Once on the watershed their only significant support base in the event of problems – for about 300 miles in any direction – was the town of Porto Velho on the Rio Madeira which had a commercial airstrip, shops and radio facilities.

The expedition started comfortably enough with the party of ten taking a cruise upstream from Manaus, initially west bound for 100 miles of the Amazon and then, turning south-west, for the 1,000 mile voyage up the Rio Purus to a small backwoods township called Lábrea.

They were a large team of botanists and INPA Brazilian field helpers (including the field man who had so nearly lost his life on the Waiká trail) and their 'cruise liner' was a small wooden river launch of around eighty tons, so they felt a little squashed sharing with ten other local passengers, animals and cargo which included, Iain noted, a ton of cement.

But despite the crowded decks Iain, as ever, managed to get some paperwork done on passage, including the framing of an application for another grant from the US National Science Foundation. It is a good indicator of Iain's complete contentment in the field that he is able to concentrate on administration in this way, though Mickey Maroncelli, his (then) secretary at the New York Botanical Garden, says she used to dread the notes coming through.

'They were always written in that awful handwriting of his, and on anything that came to hand: old field notebooks, exercise pads, backs of envelopes – anything. The only thing I never had sent to me was washroom paper. Even then, you know, I'm not so sure...

'Of course then they had usually all been jammed down the back of some field press or something and then carried through the jungle with the plants or brought back in a canoe. By the time I got them, they were usually pretty crumpled, and not so easy to interpret.'

As they headed south the team made daily forays along the riverbanks ahead of the launch using their (now standard) expedition inflatables. They had three Zodiacs, two small and one large. They would inflate them in the morning – in front of a highly appreciative audience – launch them over the side and zoom off in great style upstream to go ashore and collect for the main part of the day. Then, when the riverboat had caught up, they would go aboard for supper in the evening.

After a week of gentle cruising they arrived in Lábrea and went ashore to stay with a local missionary couple there: Fred and Zeni Orr. They kindly put all ten of them up – by slinging their hammocks out on the verandah. Fred was an Irish missionary who had great experience of the forest and worked extensively with the Jamamadi. Fred and Zeni had met Iain and Anne two years earlier during a brief flying visit they had made to Lábrea. Their chief contact with the outside world at that time was via the river one way or another, either going out by riverboat, as Iain and his party had come, or flying boat, which landed nearby on the water. Fred and Zeni occasionally went to Manaus on business at mission headquarters (the Acre Gospel Mission) to collect a few personal supplies or for Fred to preach – he is still one of the most popular preachers in the whole Amazon region.

Soon the three inflatables were again in action taking different groups of botanists to areas of the river where they could conduct research within their own interests and specializations. Five went on a half-day's travel up-river to a wide grassy savannah area known as the Puciari plain to make an ecological survey of all the vegetation in the area. This was the first time botanists had ever visited this region. Two stayed in the Lábrea area collecting Bignonias, Bignoniaceae, and the others went with Iain a hundred miles south to study further the plant culture of the Jamamadi. It was a mix of botanical area surveys, special plant collections and ethnobotanical work which was typical of the broad research style on this kind of expedition.

After a week the groups came back together and with the three inflatables loaded with ten men, stores and 400 gallons of fuel – rather more than they were supposed to carry – they set out on the next leg of the journey: up the Rio Ituxi to the mouth of the smaller Rio Curuquetê. This would lead directly up to the watershed. From there it should be a relatively short overland trek across the ridge and down to the Rio Madeira.

Iain had hoped to hire a local guide to help them at this stage but he was very surprised to find no one would go with them. Three times they approached highly recommended watermen who at first seemed very keen to help, but when told the full extent of the botanists' plans – to cross the heights above the river – each one immediately got cold feet.

'No one goes up there,' they said flatly. 'We cannot help you. We do not think it is wise for you to consider going there.'

They did not say why. But Iain had determined the route and, with nothing tangible threatening them, they decided to trust their own ability to navigate on the river, and to glean more information from local people they met along the way.

The first 250 miles to the mouth of the Rio Curuquetê was completed without incident, though Iain's heavily-laden boat, the large inflatable, was unable to plane properly along the surface of the water. This meant that it stayed low in displacement mode, wallowing along, using double the fuel and moving at a quarter of the speed they needed. Their fuel stocks were diminishing rapidly. In the end Iain decided to hitch a ride, literally, by tying the heavy Zodiac up to a passing river launch.

The launch's engine – an ancient Saab four hp single cylinder diesel – vibrated so badly that Iain complained all day of a continual itch in his nose as a result of the rattling. And it had a highly specialized starting technique – applying a blow torch to the cylinder head for fifteen minutes until it glowed red hot – before swinging over!

Their destination, where they caught up with the two faster inflatables, was the small community of Boca do Curuquetê ('Mouth of the Curuquetê') where they were able to base themselves at the home of a Brazil nut gatherer and rubber tapper for a few days, in order to do some collecting in the nearby forest.

This man and his family were some of the many thousands of 'Caboclos' in the Amazon region who live by gathering rubber latex in the dry season and Brazil nuts in the wet. These they trade, or barter, for the various commodities which local traders, who ply all the rivers, have to sell.

They are a most hospitable people, and very pleased to put up passing visitors on an informal, 'hook your hammock there, cassava and coffee for breakfast' basis. Iain and his team always expect to pay for this imposition, but the willingness of the local tappers to take in strangers at any time, without notice, is something they all learned to value. Iain says that 'their unvarying hospitality means that one can always be assured of a roof and shelter anywhere on the river, though they are usually very poor people.'

The Caboclos largely depend on the rubber and nut gathering. Both Brazil nut and rubber trees grow wild in the jungle and in particular the Brazil nut tree can be one of the most magnificent in the forest, rising to a height of fifty or sixty metres with a trunk three metres across. The nuts are produced at the top of the tree in the forest canopy, following flowering in November, after which the fruit forms. This takes about fourteen months, growing into a ball-shaped pod, known as a 'pyxidia', about 25cm in diametre, hard and woody in colour and texture. Inside each of these are a dozen or more nuts (the maximum is about twenty), shaped like segments of an orange, with the hard woody shells we know so well.

Usually when ripe the whole pod falls to the ground and then (if nut gatherers are not around) the wild agouti will move in swiftly to make a meal of them. It is thought that the agouti is the only animal in the forest with jaws strong enough to crack open a Brazil nut. And the agouti's penchant for burying these nuts in the ground is an effective method of distributing them around the forest, and ensuring their germination.

Usually the freelance gatherers go into the forest to collect the nuts in February, March and April. It is generally the rainy season when the forest lowland areas near the river are flooded. The Brazil nut trees prefer the higher ground which never floods, known throughout the Amazon as the *terra firme*. The pods are slashed open skilfully with machetes and the nuts poured out and carried in baskets or sacks to the riverbank to await the passing traders. Nut gathering can be quite dangerous. The falling fruit smashes down from canopy height like cannon-balls, which is why they avoid collecting in January and February when most fruit fall.

The dry season was the time to collect rubber. The trees (*Hevea brasiliensis*) usually occupy lower ground and are not accessible when the forest floods, but when it dries out they are ideally placed – close to the river for trading and to the riverside dwellers' homes where they must 'cure' the raw latex before selling it on.

To collect the rubber the tapper makes a trail through the forest from tree to tree. This soon becomes a well-worn track and usually includes up to 150 trees. The trees are naturally distributed throughout the forest, among all the others, within about 100–200 metres of each other. The tapper starts collecting first thing in the morning, walking his trail and cutting a diagonal slash in each rubber tree to let the latex run out. Large trees have more than one cut – perhaps up to four, and, if older, can be easily recognized by the distinctive herring-bone pattern of previous cuts up and down the trunk. The tree's latex seals the cuts naturally in about a day. Tall trees may be climbed first to make cuts higher up, and for this the tappers leave a short pole with steps cut out near the tree. Under each new cut they place a small tin to collect the rapidly exuding latex.

When the rounds are done the tappers go home to breakfast, setting out again later in the morning to re-visit each tree. They collect the latex from each tin into a larger container. The collecting tin is taken off the tree and put on a stick nearby ready for use in two days time. To tap a tree every day would quickly exhaust it.

The collected latex has to be cured immediately or it will soon 'go off'. The tapper spends the afternoon stoking up a fire which must produce a strongly acidic smoke (using wood from the *Massaranduba* tree for example) and then pouring the crude latex onto the end of a continuously turning stick placed in the rising heat and smoke. Soon this becomes a sticky white ball of cured rubber. When it weighs about ten kilos (roughly the size of a football) it is taken off, stamped with the name of the tapper and stored. When the tapper has a good number of these balls (or when the trader next calls), they are sold in return for goods.

Unfortunately these goods suffer from severe inflation when they cross the deck of the trader's boat. The riverbank dwellers find themselves forking out high prices for the goods they want – rice, beans, batteries, nails, T-shirts, saucepans – but they have little option but to buy, unless they have boats of their own and can go to the village market.

The nut and rubber trade of the Amazon rivers, while poorly rewarded and monopolized by the riverboat middlemen, is important to a great many throughout the region and offers some a chance to earn at least a little more than they can get from basic subsistence farming.

One of Iain's most interesting botanical discoveries of his time at Boca do Curuquetê was made almost by accident, when a field group returned one evening munching a nutty fruit they had found on the forest floor. Iain thought he recognized a Chrysobalanaceae in the remains of their snack, possibly one of the least collected genus in South America: *Acioa* – the only American genus of Chrysobalanaceae he had never actually seen in the field.

The next morning he went straight to the part of the forest where the others had found it. Sure enough, it was *Acioa* – a new species of it (*Acioca edulis*). It was only the fourth species of this predominantly African genus to be found in the Americas.

But now they had to head up towards the watershed, using the Rio Curuquetê as their highway and taking it as far up towards its source as possible. The river was no longer wide and rolling, but narrower and interrupted by falls and rocky rapids. Nevertheless for many miles it was navigable at speed, though their large boat would still not plane, slowing down progress considerably.

There are four major falls on the Rio Curuquetê and they collected at each along the way. At the first, about two hours up-river, they left one botanist (Paul Maas, from Utrecht in the Netherlands) in charge of a group while Iain took the rest on to the second, the Santo Antônio Falls. This, noted Iain, was 'one of the most beautiful areas I have ever visited in Amazonia – a broad twenty foot falls splashing down into a natural harbour edged with an expanse of sandy beach.'

His party immediately fell to work in this idyll, collecting, counting and observing the botanical detail. On the rocks beside the foaming waters Iain noted many hundreds of coloured butterflies and they collected many plant specimens including mosses, lichens, and rock plants during a week of hard, but very enjoyable work. Another new species was discovered: a jacaranda with deep purple trumpet-shaped flowers. As Iain had always hoped, the deep interior was giving up its secrets to his persistent researchers.

But it was to be the last pleasant week for all of them for some time to come. At the third falls, the Sâo Paulo Rapids, the large Zodiac was proving too severe an impediment to their movement so they decided to split the party the next day. One group was to continue up the Rio Curuquetê and across the watershed while the other, led by student Bill Steward, returned downstream to Lábrea with the slower boat. They would then fly across by MAF plane to meet on the other side of the Rio Madeira.

That night Iain noted in his diary that the party had been plagued by vampire bats darting down and fluttering around them at their campsite. In the end they proved so aggressive that in order to sleep the campers were forced to tie their mosquito nets tightly around themselves. It was not a good omen.

The next morning those who were to return to Lábrea made preparations to leave. Another member of the party, Alan Atchley (who had discovered the jacaranda), had been unwell for some days and that morning felt very much worse. He too would have to return to Lábrea. So they set off, taking advantage of the river flow to motor during the day and drift downstream at night to conserve fuel.

The watershed party continued up-river, carrying or 'portaging' the boats and other gear round impassable rapids. There were also large numbers of logs floating in the few remaining stretches of open water. These had to be spotted well ahead and the outboard motors stopped and lifted to pass over them, which interrupted the going, making it slow and laborious.

Iain also began to feel sick. But, thinking it was no more than a passing chill, he put his mind on the task ahead. The team pressed on steadily to reach the last of the great falls – the Republica Falls. Iain's sickness had not lifted though he hung on long enough to make a very good collection. But, at the end of the day, wholly uncharacteristically, he collapsed into his hammock without finishing sorting his collected plants. He was in a high fever.

He struggled on for two more days, helping where he could to pull the equipment and dinghies over rocks and logs, and finally lending a hand to beach them for the last time at the headwaters of the river. From there on the party would have to walk.

But he had to rest continually, and was forced to lie down during the day. He was too ill to stay in the jungle. Nothing in the medical kit seemed to touch the fever so he had to try to walk out with a companion to get himself help – there was no other option. No airlift was possible even if they had been in radio communication with the outside world. No rescue party could be summoned from Porto Velho.

By then they were very nearly at the farthest point of their journey and to go on was as easy as to go back. So the next day, with two field workers to help him, Iain set out to reach a road and get to Porto Velho. He still had a high fever and needed frequent rests to cross the watershed and get down to the Rio Madeira. They estimated that at best it would be a four-day hike. But they knew there was a remote farmhouse further down the valley, and they might try and make it as far as that for the first night's stop.

Iain couldn't carry any weight in his pack and, unable to help with navigation as his head was spinning so badly, he blindly followed the field workers. They promptly got lost. His diary picks up the tale:

'We came to a large Brazil nut gathering place full of trails and so lost the way. We wandered about for four hours and I began to feel worse. Later we discovered the trail and after another hour I could not make it any further so they put up my hammock.'

He slept there for an hour and they started out again.

'After this rest I was able to drive myself forward so we went on. After another hour I began to feel terribly weak with fever – as I had not eaten for four days. I had to sit down every ten minutes. At five in the afternoon (they had started out before eight that morning) we still did not seem to be near. After a rest I prayed for strength to walk faster.' The tropical night would descend on them within the hour.

'I got up and felt better and we were able to walk the last hour at normal fast walking pace. We arrived at 6 p.m. and I just collapsed onto my haversack.'

Fortunately the farmer was at home and took them in for the night.

So started what Iain feelingly describes as 'the worst days of my life': a nightmare of fevered walking and resting, stumbling and staggering through the jungle. At their next stop they got a mule for him to ride, and the small party was reduced to two as the other field worker returned to help the main expedition who were still struggling to cross the watershed with the rest of the equipment, plant collection and inflatable boats. Iain went on with just one man to support him: the field worker whose open dislike of Amerindians had so jeopardized the expedition on the Waiká trail. It was not a fortunate choice.

Worse was to come. The next day Iain's previously fit companion began to complain of fever and very quickly he too became sick. It looked as if he had caught the same mystery illness. Iain slid off the mule and the field worker took his place.

They continued on together crossing streams and rocks and plunging through the undergrowth, Iain's confused and feverish mind trying to work out the faint trail through the forest, frequently peering through sweat-drenched eyes at the small dancing figures on the hand compass he carried and trying to lead the mule along what seemed the safest path.

Once, crossing a fast-flowing gully, the mule stumbled. With a groan the sick field worker pitched over helplessly into the water, unconscious. Barely conscious himself, Iain managed to pull the man up and drag him ashore. Then, holding the alarmed animal steady, he lifted the dead weight of the Brazilian back up into a slumped riding position. He took off his own trouser belt and tied the man more firmly in place, wrapping the mule's bridle round him for extra security. After a short rest they started on down again.

'To this day,' says Iain, 'I still do not know how I managed to do it.'

It was four more days before they reached the dirt road to Porto Velho. A passing bus picked them up and took the exhausted pair into the town where Iain immediately went to the house of missionary Paul Bellington. Arriving there he called out 'Paul – help!' before passing out on his haversack again.

He came round to find Paul tending to his companion whose condition was worsening by the hour. Paul had seen quickly that, as they had not responded to any treatment (they had been taking drugs from the medical supplies carried on the expedition), they needed to get to hospital urgently. Iain did not seem to see the gravity of the situation and wanted to wait for the rest of the expedition to catch them up. But Paul overruled him (something, he later confessed, he would never dare to do normally) and drove them directly to the airport and put Iain on a plane to Manaus.

He sent the suffering field worker on to Belém, near the mouth of the Amazon, where his home was. He knew he would find professional help.

But there wasn't any professional help in Manaus for Iain when he arrived. It was Sunday and none of the hospital doctors could be contacted. Back in their rented house Anne made Iain as comfortable as she could and then set out to get help in the car. Drawing a blank at all of the local doctors' homes, in desperation she headed for the headquarters of the Acre Gospel Mission where they had friends. There, she feels, God took a direct hand in the matter. Fred Orr was in town and he came over immediately and quickly assessed the situation.

'There is a highly resistant strain of malaria in my part of the country,' he said. 'It is rare, but seeing where you have been working, it's possible that you have got it.'

Iain protested that all of the party had been taking anti-malaria tablets. But Fred dismissed this. The malaria strain he had come across was particularly virulent and always seemed to bypass these. He quickly took a blood sample from Iain.

'They know me at the hospital,' he said. 'I will do the tests myself. Sure, the doctors will not mind.' He smiled at them. 'Don't worry.'

Within two hours he was back, still smiling.

'Just great!' he enthused. 'Iain, you've got malaria! I am so pleased! Here, take these.' And he offered Iain some tablets as he lay in his hammock.

Pleased to have malaria was not quite how Iain felt at the time, but he could see that Fred was happy to have been quickly proved right. Iain believes that without his intervention even the experienced doctors in Manaus might have taken several more days before making a correct diagnosis. For it was Fred's local knowledge which had proved the key.

It was a life-threatening illness, according to Fred, and particularly nasty. No wonder the local Lábrea guides had not wished to go higher up the Rio Curuquetê valley. They knew there was an unpredictable invisible killer on the loose up there.

The field worker in Belém nearly died, though the diagnosis was phoned through to Belém at once. It soon transpired that the others in the party had succumbed to the malaria after Iain had left them on the watershed. Over the next week, in dribs and drabs, the remaining members of the ill-fated expedition arrived at the Prances' house. All but two had caught the fever and a number were very much sicker than Iain. Anne's front room quickly began to resemble a casualty clearing-station in wartime. Though accomplished in many aspects of jungle life, Anne was no nurse. She soon became one. Pills had to be administered, brows wiped, and meals served day and night. One of the team became so dehydrated that he had to be put on an intravenous drip which Anne had to monitor. And of course the children still had to eat, play and go to school.

'It was around that time that my appreciation of Iain's field career wore a bit thin,' she comments dryly.

Under her care they all recovered. As a consequence Iain was forced to take a seaside holiday with the family to convalesce (this time it was Fred Orr who insisted) and they had a very enjoyable time in Salinopolis where the Amazon reaches the sea.

'It was all very pleasant,' he says. 'But I couldn't wait to get back to the jungle.'

13. THE HIGHWAY

The Brazilian Airforce 'Buffalo' transport aircraft staggered noisily up into the still, humid early morning air. The sound of its engines beat back tinnily off the single iron-roofed airport building which was all the jungle city of Manaus could boast in November 1973. That and one tarmac runway, mainly used by the planes of the military government then in power. The city it served was scarcely awake. Early morning was the best time to fly to avoid the turbulence and storms in the tropical heat of the day.

The passengers on board this noisy addition to the dawn chorus were not military personnel. They were not, as was more usual, young conscript soldiers flying off for an enervating tour of duty in the backwaters of the largest river in the world – to them very much a backwater in their fast-developing, progressive country. But the government feared incursion from the north; so, almost daily, troops were moved in and out of the largest province in Brazil and its northern jungle frontier.

Yet in terms of future threat to Brazilian national policy in the Amazon region the seventeen highly peaceable civilian passengers in the plane were more to be feared than any invading soldiers. For in various different ways the information about to be gathered on their current mission would soon become a persistent thorn in the side of the administration whose airforce was even now carrying them swiftly across the jungle. Even the passengers themselves had no idea that they were about to become something of a viper in the bosom of the very authority which had loaned them their aircraft. They certainly did not appear very venomous for this was simply a group of fourteen highly enthusiastic Brazilian botany students – eight women and six men – with their three course tutors.

Sitting at the front of the plane with his two colleagues, the team leader for the trip was an Englishman: ecology expert Dr Robert Goodland. Tall, fair-haired and with a commanding presence, he had an accent, even when speaking fluent Portuguese, that sounded as though it had been honed on the playing fields of Eton. He had in fact been born in the British colony of Guiana, though at this moment he was in the employ of the New York Botanical Garden. Colonial in manner, though not in attitude, he was in his element. An acknowledged world expert on ecology generally, he was also a specialist in one area of ecology – then only an infant science – which noted and studied the environmental impact of large, usually industrial, projects: dams, opencast mines, petrochemical plants, new urban developments, and so on. At a time when many developing countries were eager to exploit their natural resources with the development grants available, Robert and a few others were beginning to ask what might be lost, as well as gained, by such activities.

Now, at the invitation of the man who sat next to him in the aircraft, he was about to take a look at the possibilities – and likely problems – of the most ambitious road building programme in the world: the 1,500-mile Trans-Amazonian Highway in northern Brazil, the first part of a project to link the Atlantic to the Pacific, coast to coast, by road through the most intractable jungle in South America. He was highly delighted to be aboard.

His senior colleague, and instigator of the trip, sported a wide bushy beard, and a shy, lopsided smile. Iain Prance was about to take yet another look around the rainforest. He too was particularly pleased to be aboard, not only because he always enjoyed a field trip to the Amazon but because recently he had had rather too little time to enjoy it despite having lived back in Manaus again with Anne and the children for over six months. His time had been taken up in setting up the new course at INPA for the Brazilian students with him on the plane. A day or two in the forest near the city was all he had managed so far this year.

But now, with his students established as 'the class of '73', he felt he could permit himself the indulgence of going with them on their first major field trip – an ecological study of the Highway – and note by the by, the opportunities it might offer for future plant collecting.

For the Trans-Amazonian Highway was literally breaking new ground right through the jungle and as a field botanist Iain was viewing the new route as a kind of super 'transect' or botanical review line. The transect was a familiar tool in field research which he had used frequently since his very first expedition in Surinam. A transect is a pathway cut directly through the jungle on a single compass bearing which runs for a number of kilometres, straight out, like a Roman road, from a chosen starting-point (usually the expedition's base camp). This ensures that all collecting along the line will be statistically representative of the flora in a chosen area.

Normally a standard distance of about five metres is explored either side of a transect which can be twenty or thirty kilometres long. That is up to about a day's march from the campsite. The camp itself will usually be near a waterway or airfield so that supplies can be brought in and specimens taken out. Deeper jungle penetration (away from rivers) has to be done by planned hiking, as Iain and Fritz Harter had done in Yanomami country two years earlier. That had been a 250km walk, but their route had been dictated mainly by winding trails already used by the local Amerindians. A straight transect line has particular statistical and cartographical value. Each plant number collected along it is recorded in the botanist's field notes with the location relative to it. This information can then be accurately transferred to a distribution map, as the origin and direction of the line is precisely known.

Today new satellite navigation systems are making all of this business much easier. A modern field botanist can plot the position of any plant to an accuracy of 100m anywhere in the world, using latitude and longitude, with the satellite 'Global Positioning System', GPS (originally invented by the US military for targeting cruise missiles), but the transect still has much operational and statistical value, even today.

Cutting transects though, is very hard work and more often than not teams of local people are employed to cut out the line ahead of the botanical party to maximise collecting time. Natural and man-made obstacles intervene unexpectedly (cliffs, streams, villages, fields) and a number of educated guesses must then be made when pre-planning the transect so that it is genuinely representative of the jungle area under investigation.

All of this meant that the straight road the Brazilian government was cutting, precisely located and surveyed, 300 metres wide right through the entire forest, with superb access, provided too good a chance to miss. To Iain and his colleagues the Trans-Amazonian Highway 'transect' was a gift from on high.

But it still needed critical assessment. The prospect of easy access to so much 'uncollected' jungle was appetizing, but disturbing reports had been coming through of drastic ecological mistakes being made in the laying of the new road. Iain felt that his party had an ideal opportunity (and in Dr Goodland a world class expert) to assess the environmental impact of the project.

The third tutor was Eduardo Lleras, a Colombian expert in plant anatomy and a PhD student of Iain's in New York. He would help with some of the field lectures, lead group discussions and do some plant collecting. His main contribution would be through the microscope on their return to Manaus. Like Iain, Eduardo confidently expected a feast of new flora to inspect and classify, possibly even some new species, plucked from the roadside as it plunged west across the continent in the direction of his own country.

But if, for different reasons, the tutors had the Highway on their minds, the students, all graduates from Brazilian universities (mostly Amazonian), were pleased simply to look down with pride on the size and richness of their homeland. As the brown and green plane spun on across the matching jungle which continuously unrolled ahead of them, they chattered excitedly among themselves. To them this was the grand adventure, the most exciting part of their two-year course so far. They were moving on to front line field work in the Amazon. This, they hoped, was where they would pursue their careers. *Ordem y Progresso* – 'Order and Progress' – proclaimed their national flag, the words emblazoned under an image of the world. And they were the living proof. They were ordering their minds in the great business of scientific progress, in Brazil for Brazil. Their botany course was new, challenging, Amazon-based and government-funded. As, for that matter, was the Trans-Amazonian Highway.

The Highway, however, had a rather higher profile. It was a much championed and lauded project throughout the whole of the country, indeed the world. A colossal undertaking which would not only open up the 'backward' Amazon region – in which they could specialize for the first time – but also provide homes and an agricultural livelihood for the many new immigrants from further south (and the drought striken north-east) currently being tempted to move to the new townships by government grants, tax incentives and glowing television advertisements. New villages, towns and cities were planned, along with massive agricultural projects, all strung out along and linked up by the Highway itself.

The lush Amazon region represents over two-thirds of Brazil's territory (extending into several other countries as well): it is the size of the entire continent of Europe. If this bold plan could solve the serious over-population and unemployment of the Mato Grosso region to the south and famine in the arid north-east, what a feat of social engineering, courageous government and effective exploitation this would be! It was hoped that upwards of three million immigrants could eventually be persuaded to settle in the region. It was a project on a scarcely imaginable scale.

It is difficult now to recall the heady days of these early phases of the Highway – when so much seemed possible at, apparently, such relatively small cost; when Brazil was applauded internationally, and the loss of 'a few trees' was considered a small price to pay for such a magnificent vision. But we must try, for only when we have recaptured some sense of the tremendous enthusiasm, the urgent pressure to move forward, the sheer optimism of the time, can we understand the huge psychological wall the first dissenting environmentalists had to surmount in order to be heard, and the persistent disinclination of so many to listen to them.

But no one on the plane had any expectation of soon being cast in the role of dissenter – even though the three tutors already had practical concerns about some details of the projects they had heard about. Collectively they knew a great deal about plant ecology and environmental factors in the Amazon region. They knew, for example, the poverty of the Amazon soil, the prevalence of plant diseases and rapacious nature of tropical insects. They also knew that high-yield farming, in all probability monocultural farming, was one of the stated keystones of the anticipated economic Amazonian miracle. So even then distant alarm bells were jangling. But they kept their peace.

Except, that is, when the bright red arrow of the new Highway shot into view under the wing some two hours into the flight. Then everyone crowded to the windows, pointing and talking. Robert Goodland began to make an ecological point or two to his students, but the airforce pilot, as if somehow sensing an opening salvo at his employer's grand project, banked sharply away to continue their flight to the airstrip at Altamira in the state of Pará, their destination.

As they flew on, Iain reflected. The bright red soil along the line of the excavated road had looked more like a fiery arrow shot into the heart of the forest, than a benign fairway along which to hunt for plants.

A killer arrow from an Amerindian bow, very long, straight and deadly, like the long arrows of the Yanomami. The thought had not occurred to him quite that way before.

On landing they were met by the enthusiastic local director of the Agricultural Research Institute, a man of consequence now that the Highway was coming through. He was bursting with enthusiasm. On his shoulders lay the local responsibility for setting up the 'agropolis' or agricultural township which had been planned as part of the chain of such population centres along the length of the road – every 200 kilometres to be precise, and they could afford to be precise, for the Highway was to be cut through trackless, townless jungle. Except of course for the tracks and townships of the Amerindians. But these were to be ignored, or pushed out of the way. The Amerindians were not to be consulted.

Every 500 kilometres a larger centre, a 'ruropolis', was planned with smaller intermediate linking villages, 'agrovillas', every 50 kilometres. In social scale it was a scheme on a par with Rameses' massive store cities in ancient Egypt or Alexander the Great's Alexandria, planted in the desert. Not since the Romans had a government been so struck with the value of long, straight roads. Some also said darkly that this showed an element of similar reasoning: the military would surely want to make use of it one day. But, for all that, it was a bold and adventurous plan. The lush jungle seemed to promise rich crops with abundant grassland for cattle and the local Institute director could afford to be enthusiastic.

An elderly Volkswagen 'combi' van, for the students, and an ancient jeep, for their tutors, was presented to the party as transport and they started out for the Highway road-head. The tired suspension on both vehicles protested in shrill disharmony as they bounced and jerked unevenly over the rough dusty track towards the construction site sending up a plume of red dust behind them.

As they slewed round a sharp bend one of the students, thirsty after the long flight, took a final swig of cola from a can and lazily lobbed the empty container into the passing jungle. Instantly a loud bellow came from the jeep behind.

'Stop!' roared Robert Goodland at the top of his voice.

The driver from the agricultural faculty was sure of one thing: someone must have died. No one shouted like that unless it was a matter of life and death. He landed hard on the brakes. The combi driver did likewise and the little convoy skidded to a hurried halt in two billowing clouds of red dust. The English ecologist clambered down from the jeep, his face livid with rage.

'Who did that?' he demanded instantly and loudly in Portuguese. For a moment the students emerging hurriedly from their van were non-plussed. Who did what? They looked inquiringly at each other.

'Who threw that can into the bush?' repeated the angry ecologist, marching off down the track as he spoke.

'This, here!' He stood over the offending article, pointing downwards. A moment or two passed and then a young Brazilian stepped forward, half-insolent, half-sheepish.

'It was me, sir,' he offered. 'But it is nothing...' He stopped, suddenly very aware that this was not what he should have said. Robert swept up the can with one angry movement and thrust it at the student.

'Take it and keep it,' he said, 'until you can dispose of it properly!

'He swung round, his concerned tones and searching glance embracing the whole group.

'*Never* forget you are the future conservators of this forest. You *alone*. For you alone will know about it, will understand properly what it is, what it represents to your country, and to the world. If you want to be botanists, ecologists and environmentalists, you must learn to behave like them. For if you do not, no one else will. Never, never, never again!' he concluded.

He turned abruptly on his heel and strode back in silence to the jeep. The students stood for a moment, the offender holding the empty can at arm's length, as though it was now infectious. Then, with a murmur, they clambered back into their combi. Iain followed Robert into the jeep and, looking back, saw a couple of them tap their foreheads and roll their eyes, mouthing '*Loco!*' – crazy. A girl giggled nervously.

'On!' commanded Robert. The driver hit the gas. 'A lot to learn, your students, Iain,' Robert commented. Iain gave a non-committal reply. He had sweated blood to get this group together and his course up and running. It was all a bit over the top he thought. But he admired it. Robert was a crusader, single-minded. He understood that – that was exactly why he had invited Robert to teach impact ecology to his students. Robert had won respect as an ecologist despite the unpopularity that his research often courted. And he was good at it. There was sound science mixed in with his passion, and that set people thinking. That was just why Iain had invited him to lead this part of the course. Robert had made a provocative unforgettable point and set the students thinking. And he was right – they *would* be their country's leading scientists in the years to come and so much rested on what they learned and believed now, before the pressures and temptations of South American professional life crowded in upon them. They would all learn a great deal about the environment. He would see to that. But would they then practise what they had learned? That was what really mattered.

And Iain himself? What did he think? Did he practise what he preached?

He knew that by now he had a good reputation as an expert in Amazon flora. His recent letters to Robert Goodland and others like him had prompted senior professional botanists to leave their lecture halls and desks to fly to Manaus and lecture Brazilian students in return for a day or two out with him in the nearby jungle. He was listened to and respected in his field.

So was it sufficient just to explore and teach in the Amazon, collect plants and then compile monographs back in New York? Or were there other things he should be looking out for, and applying himself to? 'Those who have ears to hear...' the words seemed to be a refrain in the depths of his mind. 'Or eyes to see...' he half-consciously replied. This particular journey was begining to have unexpected personal implications.

The jeep pulled up once again, the van beside it. The agricultural director got out and motioned them all forward to the low earthwork rampart ahead of them.

'We are here,' he said, almost reverently, in Portuguese.

It was the Trans-Amazonian Highway.

14. INPA

Iain's arrival at the Trans-Amazonian Highway with two colleagues and a class of Brazilian botany students had its roots in a conversation a year earlier at the headquarters of the Instituto Nacional de Pesquisas da Amazônia – INPA – in Manaus. There the new and very dynamic director of INPA, Dr Paulo Machado, had been looking for ways to improve the academic quality of his Amazonian research institute.

Set up in Manaus in 1954 by the Brazilian national government, INPA had originally been little more than a well-meaning nod in the direction of local science in the Amazon. Located in a few rooms in the centre of the city it had not been expected to do much more than look out of its collective windows and note a thing or two about the nearby forest. But a number of directors had taken their research mandate seriously and Iain, during his many visits in the sixties and early seventies, had established a close relationship with INPA down the years. Many of his best field workers were INPA employees, loaned to him for the duration of his various expeditions. Men such as José Ramos and Dionisio & Luiz Coêlho, who had become friends and colleagues on the collecting trail and who, though they had no formal training in botany, could spot and name many Amazonian plant species simply from experience.

'Having an informed local field worker – who is also a tree-climber – speeds up the collecting process tremendously,' notes Iain. 'They know right away what you want and go up and get it.'

In return, Iain had always made quite sure that INPA got a good selection of his specimens from the interior (actually a legal requirement in Brazil). He also made use of a similar agricultural organization in

Belém, near the mouth of the Amazon – the Instituto Agronomico do Norte (now CPATU) – and added considerably to their herbarium collection too. But INPA, since it was in Manaus, was naturally his most frequent port of call.

Dr Machado had been appointed as the new INPA director in 1971 and Iain soon found they shared a common vision on Amazonian research. A medical doctor by profession, Dr Machado also had a genuine love of botany and Amazonian flora. Not all the directors were so disposed – Iain remembers one or two who very much 'marked time' in the job and possessed little real interest in botanical discovery. The INPA directorship was as much a political appointment as scientific – Dr Machado later went on to become Minister of Health in the national government – so his quite genuine enthusiasm was, to Iain, extremely welcome.

Dr Machado wanted to make INPA a centre of excellence, a place where scientists came to look for information and advice on Amazon flora and fauna – a study centre in its own right, rather than a drop-in place for expeditions heading for the interior.

First he decided on a move: he carved out a carefully sited, compact new campus of the forest, on the edge of the city. But he did not carve out much.

'Let the Institute be *in* the forest,' he declared. So, instead of creating sweeping landscaped grounds and wide open vistas, he put the new buildings close to the forest edge, almost sheltering under the soft foliage. He also ensured each building had a large roof overhang to keep the interior cool and minimize the use of air-conditioning. 'Now, every day we can *see* what we should be studying,' he said. They could hear it too. On some evenings, Iain recalls, conversation over supper in the director's bungalow would be all but drowned by the crackling hiss of the crickets and cicadas, the rush of bat wings out of the darkness, or the distant croak of frogs and the rattle of palm fronds moving in the breeze. It was a location that could not help but inspire a visitor to further interest in the jungle so evidently on the doorstep. Iain and Anne became frequent visitors, both professionally and socially.

One evening they were sitting with the director in his house, listening to the forest and enjoying a cool drink of 'Guaraná', the Amazonian soft drink that tastes like a mixture of lemon juice and aniseed. Iain, Anne and the children had soon to return to New York. Iain's expeditions were over for another year and they would shortly be facing the northern winter and, for Iain, another year in the New York Botanical Garden herbarium classifying and 'working up' his Amazonian collections. The conversation turned (as it often did when Dr Machado and Iain got together) to the future of the Institute.

After a pause Dr Paulo let out a long sigh and put down his glass. 'This is good here, eh?' He commented. 'A research station in the jungle, right in the Amazon.'

His audience nodded in agreement. His new campus was an excellent base for research work. He was to be congratulated.

'Then why does no one want to work here?' he growled unexpectedly. The couple paused, taken aback.

'But there are many people here,' started Iain.

'No, no. In and out yes. But not *here*. Here for good. Or at least for some time, doing real research.' Iain and Anne looked a little embarrassed. After all, they were leaving soon themselves.

'Ah,' sighed the doctor. 'I do not mean you, my friends. It is my students; my "protégés". They come here from college – Amazonian people most of them, keen botanists – then they go to America to study for their Masters and Doctors degrees, or Europe even, and do they come back to the Amazon? No. They say: "Thank you very much for the INPA grant, Dr Paulo, it was most useful, but my new field of study requires me to work in Rio de Janiero or Brasília or on the coast of Recife." *There* they have modern equipment. There they have good pay. There they have professional respect. I had two letters this morning ...' He tailed off.

Urgently Iain pressed forward. 'But this *is* the most exciting place to work, Dr Paulo. There is nowhere with such biodiversity; nowhere ... where I could strike a transect from your front door and discover maybe half a dozen new species, in fewer than five kilometres. It is a wonderful place to do botany!'

Anne sighed. Botanically it was probably true. Iain was in full spate.

'I think, doctor, it is a matter of involvement ... if your students studied here for a bit longer ... maybe even took their Masters degree here, as post-doc students ... a course just in Amazonian Botany perhaps – then they would see the fascination of it. Then they would stay. You don't need sophisticated western techniques, just a well-trained mind and a locally developed interest ...'

He too trailed off into silence. Pointing out what Dr Machado might wish for was not helpful.

The doctor sat up in his chair. 'Iain, you are right!' He became more animated. 'And I might just be able to do something about it. Tomorrow I have the director of the National Research Council coming by to see me on his way from Miami to Brasília – he has been on a pan-American conference – could you prepare me a course outline? A two-year course for a botanical Masters degree – for Brazilians – to be taught here at INPA? I could show it to him. He is just the man to authorize such a programme!'

'We'd better get to work, then,' said Iain, nodding at Anne. The party broke up instantly, the doctor shaking hands vigorously. 'At nine, then,' he said. 'You know, it might work! Iain, what an excellent idea!'

As the hot and humid night covered the jungle, city lamps burned late in the front room of Anne and Iain's stone house. As soon as they got back Iain pulled out a sheet of paper and wrote boldly across the top 'Masters Degree In Amazonian Botany', underlined it, and wrote down the first heading: Botany 1. Then he stopped. What came next?

'Anne!' he called. She was checking the children had settled down for the night and trying not to envy them their sleep. 'Do we have any syllabus for a degree course anywhere?'

'No!' came back the brisk reply. Anne knew to the air-portable ounce what the Prance family household carried and, although there was much botanical literature, none of it was a degree course syllabus.

'It's always good to start with "Student Orientation", though,' she offered. 'It shows you know what you are doing,' she added darkly, coming into the kitchen. 'But from then on it's all up to you, professor...!'

But of course it wasn't, for Anne had her own academic background to help their discussions and the practical experience of working as a teacher in three countries: England, the USA and Brazil. So, on piles of scrap paper – sheets torn from notebooks, backs of botanical surveys, anything which came to hand – they roughed out the course together.

By four in the morning it was ready, and the Guaraná bottles empty. Iain reckons Guaraná is better than coffee at keeping the mind awake – not surprising since it contains considerably more caffeine.

Their outline was for a two-year tropical botany course based solely in the Amazon area, for students who had already been awarded a degree from a Brazilian university. It would be taught at INPA by an international team of visiting professors, who would fly in for short periods to lecture the students. Iain knew they would need a local course director but was in no doubt that Dr Machado could easily choose someone for this, if funding was approved.

Iain also suggested that many of his friends, colleagues and contacts from both the USA and Europe would be pleased to come as visiting professors for a month or so in return for a guided look around the rainforest. He himself would be delighted to help out in this way, if he were free.

Neatly written up by Anne, on sheets extracted from the children's school exercise books, the new prospectus was hurried round to the director's office by nine the next morning. Anne and Iain then went to bed.

At two o'clock that afternoon they were walking over to check the INPA plant presses when they met Dr Machado. He was a picture of contentment and delight. 'Your course is approved and funded!' he chortled.

'*My* course?' replied Iain. 'I thought it was yours, well INPA's, I mean.'

'Indeed it is,' he answered. 'But I want *you* to come back and direct it!'

And, despite many protests, that is what happened.

Dr Machado flew to New York and met the president of the NYBG and Professor Krukoff who jointly approved 'lending' Iain to INPA for the next two and a half years – the time it would take for him to see through the first course of students and induct the second. Iain was to direct the whole thing and ensure academic standards were maintained. After getting over the shock of being neatly hoist with his own petard, he accepted the post with enthusiasm.

But it came at an awkward time for Anne. The family had been under considerable financial constraint over the years (she and the girls had had to pay their own air fares to Brazil each time Iain went on expedition) but they had just, at long last, found a house in White Plains which they could afford to buy. They had made an offer, bought the house and moved.

As Anne puts it: 'The new agreement with INPA left me about three months in which to enjoy my first proper home before having to jet off down south again to the jungle – for two and a half years. I put it down to God's humour, myself. He was probably making sure I did not put my hope in my "material possessions". Mind you, I didn't feel so very happy and spiritual about it at the time!'

They sublet their new home to a Japanese family and returned to Manaus with Iain and the girls in the summer of 1973. She was to help him set up the course. Their accommodation was a small bungalow on the INPA campus.

The students were drawn from the three main Amazon Universities: Manaus, Belém and Cuiabá in Mato Grosso. There were also two students from Rio de Janiero. They had all completed their degrees, but the standard was not quite the same as that of a European or US university, so some quick rearrangement of the INPA course was required.

Early on, one of the students had stopped Anne on campus and explained that she was rather overwhelmed by the reading Iain seemed to require for the course. She pulled out a long list of books to be used for reference. 'I cannot read all of these,' she complained. Anne gently pointed out that only parts of the books needed to be read at any one time and there would be different parts to read as the course progressed. They were all standard reference works. She received an uncomprehending look. It dawned on Anne that the use of a book for reference was not something familiar to this student. Clearly she either read a book completely or not at all. So she went and asked the others. She found that they too had been very worried about the reading list but had not so far dared to say anything. Anne talked to Iain.

'You need to programme in "Student Use of the Library" before this course gets much older,' she advised. Consequently two days of 'Library Science' was quickly included plus another on the use of the Library of Congress book classification code.

The same thing happened with microscopy. It soon emerged that few of the students had ever had the chance to try out a microscope, let alone learn to use one properly. Eduardo Lleras dropped everything and took the students for a crash course of practical microscopy for a week. Then there was Iain's old bugbear: botanical Latin. He bravely took that course on himself, remembering the saying that 'there's nothing like teaching something to make sure you understand it thoroughly yourself.'

'It did improve,' he recalls, 'and I was asked to teach it again later in New York, so no doubt by then I was considered something of an expert. But that was never my opinion!'

Even with Anne's full-time help (she taught English to the students as well as helping with the course and advising them personally), Iain found directing the new coursework at INPA very demanding. There was little time to engage in field work, certainly in the early months, and even less time for taxonomy in the INPA herbarium. Though set against this was the immediacy of the forest and the enjoyment of seeing the students realize what a magnificent resource they had just outside their lecture room. This, after all, was the whole point.

He was pleased too to welcome so many of his friends from overseas to lecture the students on specialist subjects, and then to go with him on brief trips into the forest. But that just whetted his appetite for more work in the interior.

So it was with great anticipation that he had boarded the airforce plane to take three weeks away from Manaus to visit the various sites of the Trans-Amazonian Highway. He had no idea it was to prove a turning-point in his outlook on botany.

The jeep drew up at the ridge and the Volkswagen van pulled up alongside it. The party of students and three lecturers got out, following the local director of the Agricultural Institute, and walked up the earthwork. They reached the top and the mighty Highway they had seen from the air lay spread out before them. Robert Goodland looked at it long and hard. The students gathered round. His arm swept out and he unleashed a torrent of Portuguese.

Away to their right was the road, a wide red swathe cleared through the jungle and, where the trees had been felled, cut and burned, the earth had already begun to slip away down a gully.

Robert Goodland pointed. The very basis for agriculture was already fast disappearing. The students were appalled. It was so obvious. Even they, who as yet understood little about the ecology of the region, could see that if the topsoil was slipping away then there was no hope of productive farming. Did no one else see this? The director shrugged. There was other land to plant. They were clearing many many thousands of acres. If the land by the road slid away it was a small problem. Iain remembers, 'I thought that 1,500 miles of landslip would actually be quite a big problem, but I said nothing. I was rather shocked by the whole thing.'

Robert continued to outline other problems: the likely loss of nutrients as they were washed away from cleared areas when the rains came; the questionable suitability of monocultures for farming out there (at this point the director made stout defence of his new projects); the fears of a dustbowl ecology developing around the agrovilas; the loss of valuable species in the clearance areas; even the possible effects of carbon monoxide from heavy traffic.

It was quite a list, yet Iain could easily add some items to it himself. Further conversations with their guide only served to deepen his concern. What was happening before his eyes was the beginning of the rape of the rainforest. The extensive regional development plan so championed in Brasília was about to founder on the rocks of botanical ignorance. The advisers to the government were professionals, tutored in the beliefs and practices of temperate regions. They had said it was possible to farm cleared jungle and everyone believed them.

Yet Iain knew now, for a certainty, that it was all going to fail. He had known the forest now for ten years. He had spoken, and listened, to the Amerindians. He had studied the fiasco of Fordlândia.

It is difficult to imagine the overwhelming sense of horror that swept through him during that fortnight in the field – someone who knew, or could forecast, the awful truth of the situation. There was such a bitter taste to it all. The rainforest was to be relentlessly cleared, square mile upon square mile, to provide ranches, fields, towns and cities. It would bring about the certain extinction of many species and the widespread killing of plants and animals and – directly or indirectly – the extermination of many Amerindian tribespeople: it would all be to no purpose.

It could never work. He was staring at wanton destruction, and destruction of the thing he loved and had invested his life in. What, if anything, was worse was that it was all being done with very best of intentions, the very highest of motives.

Iain knew then that he had to do what he could to stop it. But he was not sure how. That first night he was taken aside by Robert Goodland. He was as concerned as Iain, though perhaps less personally involved, less personally committed to the rainforest.

'This is a disaster, Iain,' he said. 'We have to do something about it. I intend to write a paper. A book. Will you help me?' Iain agreed.

But when, a few months later, Robert Goodland wrote to Iain in Manaus outlining his thoughts for the book and asking if he would co-author it, Iain found himself in a deep quandary. The book would not pull any punches. That, Iain knew, was Robert's style. And anyway the matter was so serious he would be unhappy if it did. He wanted a big splash made, a loud whistle blown. Iain had certainly been very deeply affected by what he had seen. He had made immediate changes to his INPA course in Manaus as a consequence. Environmental issues now had top priority. He had also contacted Brasília.

But if he published a high profile book directly attacking the project head on... ? He knew he would be ejected from Brazil immediately, without appeal, and prohibited from returning. He remembered the fate of Dr Davies in Turkey. His long-standing INPA connection would also be severed and his course would be interrupted, more than likely closed down. His fourteen 'Brazilian conservators of the forest' would go their separate ways – untaught. What chance of a sound botanical input to the future in the region then? And as for himself, once placed outside the country, he would be powerless to do any more for it. He was not an ecologist like Robert. He was just a taxonomist, a front-line field worker who had suddenly realized how much he cared for the rainforest.

It was a hard decision. But in the end Iain wrote back declining the offer of co-authorship. His established work in the jungle was the instrument he believed he should use to effect change. Change from within, he hoped. He would work to save the rainforest from inside Brazil – and hope to instil in his students the concerns and cares he had for it.

Robert was profoundly disappointed, and initially very annoyed, though he understood Iain's view. But, still determined to go into print, he found another influential co-author: Howard Irwin, by now director of the New York Botanical Garden. Iain contributed just one non-controversial chapter on Amazonian vegetation. The book was called *Green Hell to Red Desert* – a vivid picture of the Amazon rainforest being laid waste – and it took the botanical world by storm.

With detailed diagrams and footnotes, it outlined the devastation to come as a consequence of the projected Amazonian developments. And the storm it bred grew and grew and soon burst out of the quiet esoteric world of plants and professors and became a whirlwind issue to governments, meat and timber importers, environmental groups and anthropologists. 'Save the Whale' soon became 'Save the Rainforest' as every Sunday magazine and television documentary seemed to focus on the plight of the trees or the Amerindians, the flora or the fauna of South America. From pop culture to academia the rainforest was on everyone's lips.

Robert Goodland was told he was no longer welcome in Brazil. Iain quietly kept going at INPA, for the moment content to hang on and wait until the storm-clouds blew over. He was convinced there would still be much work to be done to put a brake on the headlong rush to tear up the forest. Concerned western public opinion, fired by *Green Hell to Red Desert*, would stir awareness at all levels. But, though powerful, public opinion was a very fickle thing. New concerns would arise, the spotlight of publicity would move on. His would have to be the long-haul approach to saving the forest.

'I think I was right to stay put and keep quiet,' he says now. 'Different circumstances require different reactions from different people. I felt I could be more effective working there than firing salvos from abroad. Even Robert took my view later, when, as a career ecologist, he joined the staff of the World Bank – an organization little loved by ecologists generally. He told me he thought he could serve ecology better working from the ''inside''. We are still good friends. His was a vital role in alerting the world to the onset of the terrible deforestation at the time. But the battle to stop it still goes on, over twenty years later. We are both still fighting it – all over the world.'

15. CARING FOR CREATION

Iain's change of outlook following his visit to the Trans-Amazonian Highway was considerable. The visit not only marked a change in his thinking as a botanist – no longer could he be just an observer; he now had to become an active conservationist – but also the beginning of a much clearer sense of his calling as a Christian.

In his new commitment to save the rainforest – and his supportive ethnobotanical work with the Amerindians – he had found an urgent and worthwhile mission. Something over which, in the words of the famous hymn, he would have to 'wrestle and fight and pray'. Although he was in no doubt that God had called him to a career in botany, and that this was a thoroughly worthwhile business, he still had the feeling that there was something more he ought to be doing, as a man of faith. Simply pursuing science to the best of his ability was not quite enough.

Iain's profound conversion to active Christian faith at Oxford had made him aware that there would be causes he would have to fight for. And, more immediately, through the day-to-day living out of his faith – his personal Bible study, prayer and frequent fellowship with Christian missionaries – he was confronted by the calling to preach and teach the Christian message. He freely admits that the gospel message 'is everything to me'.

Preserving our world is, in his opinion, one of the most important and natural jobs human beings have been given to do. After all, he says, in Genesis, the work for the two first created people was 'to take care of the garden'.

Botany was a worthy and godly business he knew, but, deep down, he also needed to undertake a wider mission – it was a natural consequence of his faith. The battle for the rainforest – and later the continuing battle to preserve the ecological stability of so many areas of the planet – proved, for him, to be that mission.

To Iain much of the evil of humankind shows itself in the wanton destruction of nature. In stopping this, in redeeming it where he can, he has embarked on a mission for which his faith and his understanding of botanical science has well prepared him.

Not that he was pleased to have to do it. He would much rather people had quickly seen sense and left the rainforest alone (or better, researched ways of exploiting it wisely – which he now encourages). But a long-running battle had to be fought and he felt that he should be on hand to be of service in the defence of the forest as a scientist and, just as importantly, a Christian.

The visit to the Highway was also an important moment of maturity for him as a scientist, one which every scientist must face one day (much as they may wish it otherwise). He realized that his work could never be seen in isolation, as noble and detached research, apart from political or moral consequences. Science, precise rigorous observation, followed by the drawing of clear conclusions about the nature of the universe and any consequent application of this, is a powerful business. It scarcely needs underlining. But it is not divorced from the moral, spiritual and political debate which attends all power. Iain now realized this in full. And in some ways he was fortunate to find the issue so clear-cut and his moral perspective so well formed when the realization came.

None of which is to say that his efforts for conservation have been easy or comfortable. His low profile during the early days of the building of the Trans-Amazonian Highway was in contrast to some much louder campaigning later on in his career, as he wrote books, papers and seminar outlines on a great variety of Amazon issues – from recording the threatened lifestyle of the Caboclo rubber tappers to successfully lobbying for great swathes of forest to be protected and reserved. He has fought for the upholding of the rights of the Yanomami and other Amerindians whose land is even now being over run and poisoned by the random mining of the thousands of *garimperos* – gold prospectors – in Brazil. He has spoken up against the massive logging for charcoal now taking place in the Amazon and actively supports the exploitation of the large natural stands of a particular woody palm known as the babassu palm (*Orbignya phalerata*) which is far more effective as fuel (which is all that the charcoal is wanted for) and is also easily cultivated, re-planted and sustained. It's the same today as it was in the seventies: all forest cleared is forest lost.

Alongside this is his continuous and highly-regarded scientific work which has ensured that his conservation message is listened to in high places. He knows better than anyone the hard facts of the fragility of the world's ecosystem (his view is much wider now than just the Amazon) and this rightly carries great weight. Highly-charged rhetoric may achieve certain things, but speaking softly with solid scientific authority achieves much too, and the latter is very much more Iain's style.

Iain firmly believes that this work he does is of God, not in a self-aggrandizing or over-spiritual way, but simply as a believing Christian.

He believes all Christians have a biblical mandate from the Creator to care for the earth. As a scientist he is able to take it a stage further: having been given a special opportunity to investigate minutely a very significant and foundational part of God's creation, he is keenly aware of the responsibility that he now has put upon him.

His current topics of concern are wide ranging – one week he will be speaking on the importance of controlling world population, another on palm research and plantation on the island of Madagascar, now rapidly losing its rainforest. He has travelled the far reaches of the Arctic and the near reaches of the world's major inner cities, both areas where vegetation has to hang on grimly to survive, affecting the whole quality of life as it does so. In the course of his work he has visited television and radio stations the world over, spoken to hundreds in large lecture halls and answered questions in small seminar rooms, held face-to-face discussions with princes and prime ministers.

In 1994 the newest (and largest) international prize for 'work on harmony between humankind and the environment' – the International Cosmos Prize – awarded by the Expo 90 Foundation of Japan, was given to Iain, the first ever recipient.

But Iain has done his share of 'pulpit' preaching and teaching too. In Brazil, the USA and, more recently, in the United Kingdom, he has been called on to speak to congregations, church groups, schools and Christian missions about his faith.

Perhaps the most unusual occasion was one he remembers as 'one of the most thrilling experiences I've ever had'. It was, inevitably, in the Amazon, in the field – while exploring the Rio Ituxi.

He and his field team had camped for the night by the river, close to the homes of some local Caboclo families. As always, after supper had been cleared away, the plants packed into the field presses and the collecting of the day reviewed followed by planning for the next day, Iain had taken out his Bible and read quietly to himself. This was a common sight to his companions, and they thought nothing of it, as it was Iain's usual way of ending the day. But to one of the Caboclos it was a startling sight. The next day he approached Iain at the campsite.

'Are you a Christian?' he asked directly. Somewhat startled Iain asked him to repeat himself. 'Are you a Christian?' he was asked again in Portuguese. Iain confirmed that he was and the man's face lit up with a broad smile of excitement. 'Alleluia, thank you God!' he shouted in delight. Iain immediately expressed his surprise at finding someone of obvious Christian faith apparently so far from any Christian community or influence.

The man then explained that he had been first attracted to Christian thinking through listening to the broadcasts of a radio station known variously throughout the northern countries of South America as the 'Voice of the Andes' or 'HCJB' (the letters of its licenced call-sign). A missionary radio station supported by mainstream Christian churches, HCJB transmits music and Bible teaching programmes from its headquarters in Lima, Peru. He had come across one of their programmes while spinning the dial on his transistor radio one evening. The programme's subject-matter had caught his interest and he had stopped to listen. He then went on to listen more frequently and had eventually, one day on his own in his hut, made a Christian commitment.

He had then written off for a Portuguese Bible, which had been offered free by the radio. A year later one arrived, having taken a remarkably circuitous route along several rivers and across a number of mountains in its journey from the heights of Peru to the backwaters of the Rio Ituxi. But José had thanked God heartily for this as he had felt, not without reason, that it was something of a miracle his letter had got through to the radio station and that he had received a reply at all – river and mountain mail being what it is in the poorer parts of South America.

José himself had had to make a three-day canoe trip to Labrea to collect his Bible from a relative – the only one with a proper postal address! This Bible formed the mainstay of his own faith and that of one or two of his neighbours who had also become Christians through listening to the broadcasts and to José reading from his Bible. Now he had a small group who met regularly at his home for discussion, Bible study and prayer.

The man asked Iain if he would lead a service of worship and preach for them the next day, Sunday. They had never actually had a service, or a preacher for that matter, on the Rio Ituxi. Their usual form of Sunday worship was to listen to HCJB.

Iain was only too happy to accept such an invitation and explained to the other botanists and field workers that their planned early morning departure would now have to be delayed a few hours due to unexpected church business!

Though surprised, they were more than happy to accept this. The area had already proved very fruitful for specimen collecting and no doubt would yield still more to those who did not want to listen to their team leader preach. That afternoon Iain spent some considerable time in prayer and the preparation of his text for the coming day and then helped with the packing up of the campsite.

Early the next morning José commanded his children to run round the little community with messages urging his friends and neighbours to 'come to our very first Christian service' and to 'hear a real preacher for the first time – not on the radio!' This was entertainment too good to miss and thirty-six neighbours (most of the community) turned up to pack the main room and veranda of José's small wooden home by the river. Then he and Iain joined together in leading the very first service of Christian worship on the banks of the Ituxi river.

But although the service on the riverbank was unique in its circumstances, Iain was by this time quite experienced at preaching and teaching in Portuguese and had already spoken quite often in Manaus, where the family went to church, and in other towns and cities of the Amazon area when he happened to be there on botanical business. He does not regard himself as a regular preacher but in most cases he will accept what invitations he can, when asked, particularly in Brazil. He is fully aware of the great need for Bible teaching there (where the church is growing rapidly).

He remembers with a smile that in the early days he was often introduced to congregations as representing the 'Botanical Christian Mission' or some similar nonsense – even in the big city churches he found they were quite unable to take in that he was not in the Amazon as a missionary. They could not imagine that a scientist could be a Christian. Their only concept of a foreign visitor to a church was of a missionary.

Perhaps his most demanding time as a speaker in Brazil was when he was asked to lead a series of Bible studies on 'Creation and Evolution' at their church in Manaus.

He was called in by the minister of the church there, an energetic Brazilian, trained in Texas by the Southern Baptist denomination in the USA. This church, the first Baptist church of Manaus, was the home church for all the Prance family while they stayed in the city. The minister was very concerned as a number of his older youth group members had started to drift away from the church, having formerly been keen members. It had happened when they began to attend college.

He had talked to some of them and discovered that the college lecturers had ridiculed their faith and particularly their belief that a superior spirit could have had anything to do with creating the world. Evolution out of nothing was the accepted credo of the modern scientist or engineer, they had been told, and if the new students wanted to learn things properly they had to give up their primitive ideas right away. A lot of them had taken this to heart and, having no information to hand to counter it, had quickly lost headway in their faith.

Iain explained that there was absolutely no conflict in his mind between the Christan faith, science and evolution, and furthermore he felt that believing in God rather helped to pursue science properly. In fact many would argue that the scientific approach could only have been developed – as it had been – by those who believed in an orderly, God-created universe. The scientific assumption that the universe makes sense was, he said, one of faith.

The minister was impressed by Iain's comments and felt that a talk at church along these lines would give the young people a chance to think through the issues before facing the prejudices of their college. They could then view matters more honestly from an informed Christian perspective.

Iain offered to take a course for the young people of the church on evolution and creation – the biblical view of a Christian and a scientist, a complete study of the subject.

It is to the minister's credit – and an indication of the respect in which he held Iain – that he agreed, as many from the very conservative

Southern Baptist denomination would find a view of creation that included any element of evolution very hard to swallow indeed. But there was no doubting Iain's biblical faith and excellent credentials, so he bravely gave it the green light. Iain and another colleague, Professor Warwick Kerr, director of INPA and one of Brazil's leading experts on bee genetics, taught the course of ten weeks to a large number of young people from the church, drawing out the themes of creation from the Old and New Testaments and then explaining the scientific usefulness of the 'tool' of evolution as a theory. Not evolution out of nothing, but evolution as a creative activity employed by God.

Iain is very clear that the 'fitting' of plants into their environment and the interaction of animals, insects and plants with one another within their habitats is a matter designed by God, as an evolutionary process. He belives there is a significant mechanism through which living creatures adjust to changing circumstances and conditions by adapting themselves in order to survive. And created life forms are all in the business of surviving. He would nevertheless argue very strongly that something as complex as the leaf system of a plant, the eye of a mammal or the subtle, intricate, co-operative habitat of the rainforest could not happen by accident. Accidents tend to disassemble and degrade, not construct or build up. He believes that a very great master craftsman must be behind the move from the simple to the complex – and the splendour and complexity of the forest and all that lives in it is something he praises God for.

But when it comes to researching plant families and to deducing the particular shape of a flower or fruit then he will begin to look for the needs the plant has to survive: water, light, nutrients and so on. Then the detective trail begins with the concept of evolution as a valuable tool in the scientist's hand, with the question of why and how a plant is adapted to its habitat in order to survive.

The Amazon Calabash (*Crescentia cujete*) is a good example of this. This is a large round gourd, about the size of a soccer football, which is common in the riverside forest of Amazonia, and when harvested and cut open provides a very durable container for almost anything and everything – from canoe balers and water pots to manioc bowls – even sun-hats!

This, when small, is a soft green fruit. As with most soft fruits in the jungle, it is very prone to attack by animals and birds. But it has evolved a very unusual way of protecting itself. On the surface of the fruit small glands secrete sugar-rich nectar which has a particularly strong appeal for some very aggressive stinging ants. These ants spend their days swarming over the fruit and enjoying the nectar. They defend their source of food against anything else, driving it off with their stings if necessary. The jungle population soon gets the message and leaves the great majority of the young Calabash fruit to grow unharmed.

When the fruit is full-grown the outside becomes tough and woody. At this time the nectar dries up and the protecting ants disappear. But by now they are not needed, for the mature Calabash can take care of itself. It is a neat process and shows how the plant has evolved to meet the conditions of the forest around it in order to survive.

Iain would not argue that the plant evolved this process by thinking things through for itself or accidentally developing nectar spots which luckily attracted just the right sort of ants. To him these things are under the creative hand of God, but an evolutionary process nevertheless.

Since leading the course for the young people in the church in Manaus Iain has had considerable opportunity to develop further his view of creation theology (he is often asked about it). Much thoughtful work has been done by him and a number of other senior scientists from around the world who are Christians. Once during a prayer meeting in New York held by the American Scientific Affiliation, an association of scientist Christians (of which he is longstanding member), Iain found himself invited to lead a seminar that led to contact with the Au Sable Institute for Environmental Studies, in Michigan, USA. This, he discovered, was a centre dedicated solely to examining the appropriate Christian response to the current threats to the environment. Iain subsequently took part in one of their annual forums, and later co-ordinated one of his own on 'missionary earthkeeping'. He is now a member of the board of trustees of the Au Sable Institute. Much of Iain's deeper thinking on creative environmental theology has come out of his work in preparing and listening to seminars at this unique facility. Far from being a hothouse or ivory tower, the institute is run by practical people with a Christian concern for the environment. Much of Iain's deeper thinking in this area has come out of his work in preparing and listening to seminars at this unique facility.

Some of this thinking is summed up in a recent paper which drew together his thoughts on the Amazon area:

'The contrast has been made between the rich who think nothing of destroying the land and who create ecologically wasteful systems, and the poor who when they are allowed to have the land manage highly productive plots. We must now ask why the church has so often sided with the rich rather than the poor in spite of the many biblical reasons for better stewardship of creation and justice for all peoples. Much of the agricultural land in the Third World is controlled by wealthy landowners and large companies who practise monoculture and employ the techniques that are basically western and were developed in the northern temperate region. These techniques are susceptible to disease, predators and drought in the tropics. By contrast local people around the tropics have developed methods of farming that reduce the effects of disease and predators through the use of many crops, maintain genetic diversity through the use of many varieties of each crop and use combinations of plants adapted to the eventualities of the weather such as drought. The first system is motivated by greed and short-term profit and the second by a love, care and understanding of the land and respect for the integrity of creation. The rich want to displace the poor with their temporary systems that are destructive and unsustainable.

'Many peasant farmers in Brazil and also rubber tappers who have resisted eviction from their small farms or the forest where they tap rubber trees have been assassinated by the wealthy landowners. The leader of the rubber tappers in Acre State in Brazil, Francisco (Chico) Mendes, was gunned down for seeking to conserve the forest so that he and other tappers could make their livelihood from the forest by tapping rubber and gathering Brazil nuts. He did not die in vain because extractivist reserves are now being set up in several areas of Amazonian where the tappers are allowed to work but the forest is not felled. The rubber tappers movement is greatly helped by the local churches. To attend one of their meetings and see them studying scripture and praying what to do about their desperate situation is a moving experience.'

Iain believes that much of what is written in the Bible has direct relevance to conservationists today. He quotes, in one recent discussion paper, from the Old Testament book of Isaiah: 'The earth mourns and withers, the world languishes and withers, the heavens languish together with the earth, the earth lies polluted under its inhabitants; for they have transgressed the laws, violated the statutes, broken the everlasting covenant. Therefore a curse devours the earth, and its inhabitants suffer...'

Chilling stuff, but startlingly accurate when put into the context of the modern environmental crisis, brought about by our 'transgression of the laws' of nature. Fortunately Iain's convictions take him a stage further than doom and gloom. A favourite quote of his is from the Gospel of John in the New Testament: 'God so loved the world that he gave his only Son so that anyone who believes in him should not perish but have eternal life.' In this Iain sees both comfort and challenge. Comfort in the knowledge that God is intimately concerned about the fate of the world – and humankind – and challenge that Christian believers should see themselves as leaders in the fight against environmental degradation and exploitation of fellow human beings.

In his paper he continues: 'This will involve seeking to promote small-scale projects of appropriate technology that grow out of local cultures. The Christian can learn and respect local cultures and their use of land without compromising belief. The challenge to us [Christians] is not only to learn when and where to apply appropriate technology but to be deeply grounded in the biblical teaching about the land and about creation. We then become Christian earth-keepers.'

He concludes this discussion with a powerful illustration of the work of such 'Christian earth-keepers' in action.

The Aymara people of the Altiplano of Bolivia live in the highland of the Andes and have developed farming to include many local crops such as quinoa, amaranth, spinach, grains, potatoes and other tubers of many varieties. Each family owns a plot of land but the village elders decide on what is planted and where, each year. Their decision is based on long experience and they will tell each farmer that year to plant up with quinoa or potatoes or grain. Some years they will tell him to plant nothing. But pleased though he might be to have a quiet time of it he is not afraid that he or his family will starve. The others in the village will supply his needs. It is in fact a traditional system of letting the land go fallow, letting it rest. But the Bolivian government tried to interfere with the system and insisted that they stopped 'wasting their land'. They issued and enforced instructions to plant up all the fields. Within a very few years the system began to crash as the fields began to become unproductive.

Fortunately an inter-denominational group of missionaries working with the Aymara had read in their Bibles part of the book of Leviticus where it says: 'Six years you shall sow your field and you shall prune your vineyard and gather in its fruits; but the seventh year there shall be a Sabbath of solemn rest for the land, a Sabbath to the Lord.' There is a similar chapter in the book of Exodus which concludes: 'let it rest and lie fallow.... You shall do likewise with your vineyard and with your olive orchard.'

They immediately went to see the government department of agriculture and explained the situation. After some discussion and the clear evidence presented to them that the agriculture was collapsing, the government agreed to permit the Aymara to return to their traditional system. Gradually productivity climbed back up and things returned to normal.

This example of the everyday outworkings of a biblical understanding of how the earth should be treated is a good example of the marriage of theology and practice that so excites Iain. The mission and joy of Iain's life is studying God's creation, caring for it, and urging others to do so.

16. NAMING THE LILY

Re-created under glass in the modern and extensive Princess of Wales conservatory at Kew are the three major warm growing environments of the world: tropical, sub-tropical, and desert. The interested visitor will pause at many different places in this new high-tech, eco-friendly hothouse. There's plenty there: many strange and exotic plants to catch the eye, and the evocative atmosphere of the simulated rainforest (complete with its own rain). But there is one place everyone stops: the tropical lily pond. Modern visitors are every bit as fascinated as their Victorian predecessors by the largest water-lily in the world: *Victoria amazonica*, which grows here.

Its giant round lily-pads with their distinctive upturned rims look like great green frying-pans. They float solidly on the water surface, each one a metre and a half across, just touching the tip of its leaf against its neighbour and forming a pattern like counters on a giant draught-board. Each pad is buoyant enough to support a small child and at the centre from where the pads emerge is a remarkable large, lotus-like flower which changes colour from brilliant white to deep-reddish purple. As its name implies it is a queen of the Amazon.

In 1838 Sir Joseph Paxton, gardener to the Duke of Devonshire (and incidentally designer of Crystal Palace), was the first man to flower this lily in Britain. He was instantly co-opted onto a committee which was about to decide whether Kew should be taken out of the royal estate and given to the nation, or whether the garden and plant collections should be broken up. They advised (and fought for) the former, and thankfully succeeded, though it remained under royal patronage. Sir Joseph, rather like Iain, was clearly a man of many dimensions. But it is for his work with this giant South American water lily that he is chiefly remembered. It proved little short of a sensation when introduced to the gardens, even among the sober-minded Victorians. Etchings of the plant and its constituent parts soon adorned many nineteenth-century drawing room walls.

Early in 1974 Iain took his INPA students on a field trip near Manaus to study how various aquatic plants were adapted by evolution to fit their unusual environment. Included in this tour was a special look at *Victoria amazonica* – though it was not something easily missed – and the impressive plant soon prompted many questions from his students. But Iain began to flounder with his answers. It seemed to him that the more earnest their inquiries, the blander his replies became.

He cut open a closed lily flower and showed them the dozen or so beetles crowded inside. 'Why are they there?' asked the students. 'Pollination,' came the reply. That was simple enough: cantharophily – the phenomenon of beetle pollination – was well established and understood. 'How did they achieve it?' Ah, that was a bit more difficult. They rubbed up against the pollen-heavy stamens and anthers of the flower no doubt, and then somehow carried it with them, at the right time, onto the style of another flower. He did not know at what stage of the pollination cycle, nor did he know exactly how they carried the pollen (being slick, shiny creatures, not furry, like bees). All he could say was that there were a number of possibilities.

'When do they do it?' persisted the students. No answer. 'How long does it take?' No answer. 'How did they know when to do whatever it was that they did?' No answer. Iain soon began to feel it was high time to move the class on to another part of the river.

But their questions stayed with him and Iain (who by then was itching to do some practical research as well as lecturing) decided to conduct a botanical investigation into what was to prove the highly exotic sex life of this mysterious aquatic giant.

The lily grew on river backwaters and lakes close to Manaus and preparatory investigation showed that much of the scientific study must be done in the evening and at night. This fitted in well with Iain's daytime lecturing and directing schedule. He could teach by day and then stand around chest deep in muddy water all night probing the secrets of the lily pools. Sleep would come a poor third. When Anne heard about this she put her local washerwoman on immediate standby.

So Iain recruited the INPA entomologist Dr Jorge Arias and a Kew botany student Malcolm Leopard (now curator of the Zimbabwe Botanic Garden) and they started work, using an inflatable boat to get among the lilies or wading out deep into the murky water themselves with their thermometers, rulers and cameras night after night to make observations and compile their data. Their eventual findings were truly remarkable and neatly illustrate the amazing complexity – and incredible precision – of ecological inter-dependence in the rainforest.

To understand the pollination cycle of *Victoria amazonica* it is important to have a mental picture of the actual flower within which so much activity takes place. It is best to imagine it open, floating just above the surface of the water. It is a large flower, about one third of a metre in diameter.

The outside has four large sepals, or outer covers, overlaid with sharp spines. When these open the spines project down towards the water and, along with spines on the flower stem and those on the underside of the lily pad, form a formidable defence system against marauding predators. Iain says he always felt relatively safe from passing water creatures when surrounded by lily pads, though he had to watch he did not catch a finger on one of the spines himself! On some nights he did find, reflecting the beam of his flashlight, the bright glint of eyes belonging to cayman alligators just peeping above the water; though in the event they never caused him (or the rubber boat) any trouble.

Inside the outer sepals lie the many initially white petals which, when opened, make up the beautiful and distinctive lotus shape. Inside these lies a circle of thicker projections called staminodes (sometimes known as sterile stamens) and then, forming another ring further inside the stamens. The staminodes number between 14 and 35 in each flower, the stamens much more: between 104 and 330. These carry the all-important pollen on long anthers which extend into the middle of the short tube they form in the flower head. Anything, such as a beetle, that gets into the centre of the flower, cannot but emerge covered in pollen, though Iain was keen to discover exactly how it would stick onto the beetle.

At the bottom of what therefore amounts to a small tunnel leading vertically down into the centre of the upturned flower comes a further ring of staminodes. These, as they open and close, act rather like the lower waterproof hatch at the bottom of a submarine conning-tower. They are the inner seal above a large empty chamber in the centre of the flower head. And the outer staminodes act rather like the upper hatch, sealing off the 'tunnel' at the top.

At the base of the large, empty chamber is the style, the female part of the flower (the stamen is the male). All around this is a veritable larder of starch and sugar, located in paracarpels which lie in a circle around the top of the chamber. It is a complex arrangement, but as Iain found, every part of the flower has its role in the lily's pollination cycle.

This begins when the tightly-closed flower bud rises to the surface of the water before it is ready to open, just showing the green tip of the tightly-closed sepals above the water level. The bud has grown up from the lake bed towards the daylight which shines through the small area of clear water between adjacent lily pads. This is the start of the first day of the cycle; and the difference between day and night on the flower's development is very important.

Throughout the first day the spiky bud continues to push its way up above the water surface. By the end of the next day the bud is completely above the water and in its flowering position. But it is still closed. It is waiting for something...

At about six-thirty (in the latitude of Manaus) sunset occurs. Incredibly, the waiting bud is able to tell that the light is fading and, in a manner which all who have seen it say is wonderfully mysterious and magical, the flower begins to open. Like curled up ballet dancers on an empty stage the new flowers seem suddenly to spread out their skirts and rise to dance as the night begins.

The opening movement is fast – for a flower – almost visible to the naked eye, and Iain confesses that the sight of a lake full of lily flowers slowly unfolding in the half-light of dusk is one of the most breathtaking he has ever seen. The flowers come out with the stars.

The sepals open first, to an angle of 45 degrees, followed by the white petals, and then continue on to become horizontal. The petals remain mainly upright and, being a radiant white, stand out in whatever light there is (and there may be quite a bit if the moon rises) forming a sharp contrast to the dark, shiny lily pads and the still, black water. Then, with a flashlight, the small tunnel into the centre of the flower, through the tight circle of staminodes, can be made out.

Earlier, during the afternoon, a sweet fruity fragrance has been manufactured by the flower and has been building up inside the closed bud.

This becomes vaguely apparent towards the end of the day, but wafts across the lake, heady and powerful, when the flowers actually open.

This strong scent is an invitation to certain scarab beetles to come over for an all-night party. They merely have to follow their sensors to find the meeting place. And hundreds of them do. The flying beetles descend in droves on the newly-open flowers and immediately squeeze themselves down the central tube to the inner chamber which is full of good food – in the sugar- and starch-rich paracarpels. Here they begin to enjoy their evening meal. It is also pleasantly warm inside – metabolic reactions taking place in the flower raise the internal air temperature to exactly 11 degrees (Celsius) above the ambient air temperature outside – a very good place to be if you are a scarab beetle.

Iain and his colleagues found that four types of beetle readily accepted the invitation to dinner. By far the most common (90 per cent) was a variety they initially assumed to be a common scarab, *Cyclocephala castanea*, but later proved to be a completely new beetle species now named *Cyclocephala hardyi*. He comments that it would be a highly interesting piece of parallel research if an entomologist could study this plant–beetle interaction from the point of view of the life cycle of the beetle.

The evening wears on with the beetles feeding happily, but then at about midnight, an apparently sinister transformation comes over this happy scene of feasting amid the flowers. For, slowly, the flowers all begin to close. This starts with the flower's paracarpels shutting off the central access tunnel, sealing all the beetles inside. The beetles are now completely trapped.

Iain's team observed that the optimum number of beetles for a comfortable time at this stage was between six and eight, though many more can cram in if there is nowhere else to go. The highest number they counted in one flower was forty-seven! The food ran out early – if too large a number of beetles is trapped in this way this happens and then they start to damage the inside of the flower. In fact it was observed that if more than six beetles were trapped then some internal damage always occurred.

Between one and two hours after the flower closes (the scent has also much diminished by now) a colour change begins to creep over the petals. Starting from the inside and working out they begin to blush a soft pink and, as they do so, the air temperature inside the flower chamber begins slowly to fall to that of the outside air.

At dawn the flowers close up completely, staying closed for the first part of the next day. The petals continue to change hue, finally becoming a royal purple-red colour most suitable for a queen among plants. By mid-afternoon the sepals and petals re-open, the better to display the soft rich purple they have now become, but the staminodes and paracarpels, the 'lids' at the top and bottom of the tunnel, stay tightly closed over the beetles inside.

At dusk the paracarpels and staminodes relax. The sealed tunnel is re-opened and the beetles, having by now been shut up for eighteen hours, are eager to escape and so take their first chance to leave. But even in nature there is no such thing as a free lunch, so, on their way out, the beetles have an unconscious job to do. As they scuttle free, sticky with the juices they have been gorging, they pass through the circles of waving stamens in the tunnel whose anthers by now are busy dispensing pollen for all they are worth. A great deal of it attaches to the sticky beetles.

It is now sunset and, nearby, new lily flowers are opening white petals to the approaching night. Their sweet fragrance is carried on the breeze and the prospect of another feast lies before the pollen-dusted beetles. They fly to a new white flower and burrow down to the central chamber, at the bottom of which is the flower's stigma ready to receive pollen. The jostling insects rub against it and the pollen is deposited, quickly being transmitted to the flower's ovaries and triggering the release of its own pollen twenty-four hours later, just in time for the beetles to carry it onward in their next flight.

And so the complex, elegant and very intimate pollination cycle of the giant water lily is complete. The beetles are fed and the lily has its gene pattern conveyed to another in return, enabling this majestic giant to continue to grow and expand in regal splendour across the lake. A truly amazing and precise sequence of insect–plant co-operation.

These fascinating results, accompanied by diagrams, notes, graphs and charts, formed part of Iain's well-received paper 'A study of the Floral Biology of *Victoria amazonica*' published in 1975 in English in the Brazilian botanical magazine *Acta Amazonica*.

But behind the clear observations lay many murky nights with flashlight, machete and rubber dinghy. Anne remembers early mornings when, just as dawn was breaking, she would look sleepily over her hammock to see a large and odiferous collection of soggy lily plants make its way into the living-room of the house, carried by an exultant Iain delighted with the night's adventures, and whose soggy clothing was virtually indistinguishable from the plants, except perhaps that the clothes were possibly a little muddier.

'Then for Iain it would be a quick shower, on with shirt and slacks and in with breakfast, then back out for lectures. A little later in the day he would call for the plants.'

Lily flowers were collected at all stages of the cycle, from first rising bud to final, collapsing, pollinated flower. Photographs were taken and the cycle itself watched many times, with careful notes made of flower dimensions, hour of day or night and internal flower temperature. A thoroughly detailed picture was built up. It was a unique opportunity which Iain seized with characteristic energy and vigour – after 1975 he was never again to spend so much time in Amazonia. And of course the study was wholly appropriate for a future director of Kew – for today the chief executive of the Royal Gardens knows more about its chief tropical attraction than anyone else.

There was another paper Iain wrote following this research which caused a minor furore when published. It was called '*Victoria amazonica ou Victoria regia*' also printed in *Acta Amazonica*. This was a discussion on the right name to give the lily – according to the international rules of plant nomenclature. These state that the oldest species name must be used, whatever the genus, which is what Iain had done. But others, particularly in Brazil, thought otherwise.

The lily was first described in 1836, from material collected in Brazil, and named *Euryale amazonica*. But two years later it was re-described from material collected in Guyana, and called *Victoria regia* by a loyal British subject, Dr John Lindley (who later served on the same 1838 Kew committee as Sir Joseph Paxton). Because it was realized that the lily did not belong to the genus *Euryale*, but was indeed a direct genus, technically it now became *Victoria amazonica*, under the rules.

But the name *Victoria regia* had been kept in Brazil. When Iain wrote up his findings in Portuguese, pointing out the correct name, a Manaus newspaper picked up the story and accused Iain of metaphorically taking 'their' plant by giving it another name. In fact feelings ran so high that the 'Manaus Academy of Letters' – an august and authoritative body in all matters of Brazilian text – put the matter to the vote at their next committee meeting, and passed a majority for the name *Regia amazonica*!

Unfortunately for these enthusiastic men of letters, and Brazilian pride, Iain not only knew his botanical Latin, but had interpreted the rules correctly, so *Victoria amazonica* it stayed. But don't mention it in

Manaus, for they are still rather prickly on the subject of their spiky water lily.

Over the years Iain has discovered and named or re-named a fair number of new species. This is sometimes in the field, but more usually in the herbarium laboratory, where detailed examination of new specimens is carried out and queries can be quickly resolved by comparing new plants with specimens of others already collected and identified in the genus (held in the herbarium library) or by reference to plant monographs.

It is a standing joke amongst botanists that real taxonomists can only recognize a plant when it has been cut, dried and put in a herbarium reference folder. In the field, where everything is green and living, they supposedly recognize nothing. Iain, a dedicated field as well as laboratory scientist, takes such a jibe as a professional affront – but also confesses he has known one or two taxonomists who did rather tend towards this cloistered stereotype. He admits that he did once collect a specimen of *Dichapetilum*, a genus on which he specializes, and only recognized it as such when it came out of the plant press! Nevertheless it is usually in the herbarium that the new 'finds' are made.

This produces a rather curious situation where many new species have been collected in the field by people who, at the time, had no idea that they had found anything special. They may have had a suspicion that they had come across something new, and collected it on a 'hunch', but until the particular plant is cut, pressed, numbered and sent off to the right expert for proper identification they do not know.

In fact most new specimens likely to be studied in the herbarium have been sent in to be identified by a specialist botanist. For it is usually only the expert on a certain plant family who can assure the world that a new species really has been found. These people are spread throughout the botanical gardens and universities of the world, so there is a continuous international flow of herbarium specimens as different experts receive parcels of field findings to identify.

Iain makes a point of turning his specialist queries round as quickly as he can (he is expert on the plant families Chrysobalanaceae, Lecythidaceae, Dichapetalaceae and Caryocaraceae for example). Keen field collectors will be anxious to know if they have found something of botanical interest, or made botanical history. Some experts, sadly, are rather slow and so new specimens may remain unclassified for many years before being looked at.

There are two ways that plants may get new names. One is by finding a plant is new to science – a new species. The other is by re-classification. In the first case the herbarium researchers usually finds they are looking at another 'version' of a plant from a family they are familiar with. Most likely from one of their own specialist families. They can quickly tell the specimen is substantially the same as the others, but is just that little bit different. It falls into no other group – it has all the characteristics which make it one of the family, but an extra something (or lack of it) sets it apart from the previously described species.

After careful analysis (which includes accounting for unusual growth patterns due to region, altitude or other environmental factors, hence the crucial importance of field notes), the botanist will decide that a new species, or even in rare cases a whole genus, has been identified. This must then be written up in botanical Latin with an outline of the different characteristics, usually then amplified in the botanist's local language. This, when published in publications such as *Taxon*, *Brittonia* or the *Kew Bulletin* effectively puts the plant on the map and then it officially exists – with a new name. This is known formally as 'describing' the plant and it is the one who describes it – the specialist discoverer – who gives the plant its name.

It is perfectly possible when this is done that the botanist has no idea who the person is who collected and sent in the intriguing and unique specimen. All they have is their name, the location of the find and the field notes supplied with the plant.

The other way of naming a new plant species is by splitting original classifications into further sub-classifications, after identifying differences in known plants that no one else has noticed. This will happen when an existing family is researched in detail (perhaps for the first time) and it becomes apparent that several of the plants generally assigned to it simply don't 'fit'. Iain did just this kind of thing when he looked at the Chrysobalanaceae in his early days at Oxford. After some work – and usually discussion with knowledgable colleagues – it is decided to hive off the disparate group and name it as a new genus or new species depending on the extent and nature of the discovered differences.

This is the very stuff of good taxonomy, and produces a number of newly-named (though already known) plants. Once again it is the scientist, the taxonomist, who will describe the species or genus and choose the name. A good example of this for Iain was the tropical

South American bush *Rhabdodendron* which he split off from the Chrysobalanaceae. His investigations showed it to have very little in common with most Chrysobalanaceae except that the style came out of the base of the ovary – an unusual feature peculiar to the Chrysobalanaceae. But he felt this insufficient to keep it in the family and so split it off to its own family. By the laws of nomenclature the old species now belonged to the family of Rhabdodendraceae.

Other discoveries may permit a more personal choice of names and in this the botanist has an opportunity to honour, in a most subtle way, a highly esteemed colleague or friend. For while it is not at all appropriate to name a new plant after oneself, the identifier, it is very appropriate to name it after a friend, fellow scientist or someone who particularly contributed to the finding and collecting of the plant in the first place. Iain himself has been honoured by colleagues in this way – with forty-three species and one genus named after him – identified mainly as *prancei*. For example *Lecythis prancei*, a new Brazil nut, named after him by his co-worker on a Brazil nut research project – Scott Mori from the New York Botanical Garden, now senior curator there; and *Gleasonia prancei*, a plant from a later expedition to the Araca Mountains in 1986, given by Brian Boom, another colleague from New York, now vice-president for botanical sciences (the post Iain eventually held) at the Botanical Garden.

One of the people whom Iain himself has named a plant after is his Amazon missionary friend Paul Bellington. Paul once directed Iain to a new part of the forest where he felt there would be a large number of interesting plants for him to collect. Iain was so pleased when this area yielded good results, including a number of new species, that he felt it only right to honour his friend with the name of one of them, a new Chrysobalanaceae, *Licania bellingtonii*.

At the end of 1975 Iain's directorship of the degree course at INPA officially terminated and other work called him back to New York. He left with regret – the course had been a great success and naturally he wanted to follow the whole thing through once again with the new batch of students he had inducted. But Dr Warwick Kerr, the new director of INPA, was a wise and competent man who shared Iain's vision for the institute. His aim was to ensure that the course would continue. With the additional academic staff that had joined INPA since 1973 he was more than able to run the course himself on site – where it still runs today.

Iain would also miss the jungle on his doorstep – although he would return to INPA many times, he would never live there again. He would also miss the people. The Brazilian academics, people like William Rodrigues and Marlene da Silva, field workers (particularly José Ramos), and students he had come to know and respect – and the place itself, the most forward of forward bases for tropical field work. A place with an attitude to botany entirely its own.

His most vivid memory of the institute itself was the night when the plant presses caught fire. For him the memory of this dangerous and ridiculous incident, reminds him of the best and worst of the unique National Amazonian Research Institute. Fires in plant presses, while by no means common, were a very real danger in the field – with kerosene stoves burning all night, drying combustible material, often under a thatch or canvas shelter. They needed regular attention and, after a long day of collecting, a minor mistake in loading the presses or a careless adjustment made to a stove could lay the powder-train for a night-time disaster after the team went to sleep. On one river expedition Iain and his team were awakened suddenly in the night – they were sleeping on a moored riverboat – by the whoosh of a kerosene stove exploding under a plant press. Fortunately these were ashore on a near-by sandy beach. As the flames lit up the night, José Ramos, an INPA field worker and long-time friend of Iain's, jumped ashore to save the day by kicking the violently burning stove out of the way and throwing sand on the presses.

'His quick thinking saved many of the plants – and the presses – or the expedition would have been a washout,' Iain recalls. 'When it was all over we cleared up the mess and went back to our hammocks thanking God very heartily that we had put the stoves on the beach and not set them up on the boat. I dread to think what would have happened if we had!'

In Manaus, instead of portable plant presses run by kerosene, INPA proudly possessed a large electric oven for drying out the collected herbarium specimens. This, housed in the preparation room in the botany building, not only boasted electric heating but electric thermostatic control as well – which meant that the oven could be left unsupervised most of the time (even modest overheating in a plant press tends to burn specimens to a blackened crisp). But not quite trusting the thermostat (and being aware of the irregularity of power supplies in Manaus) it was institute policy to shut down the oven last thing at night, for safety's sake.

On the night of the fire Iain had this responsibility and at about 10 p.m. wandered over to turn the oven off. It was full of new material as he had just returned from a two-week expedition. As he approached the preparation room he saw wisps of smoke curling up around the half-open door. Thoroughly alarmed he ran the last few steps to the doorway and, ducking inside, was forced to gulp for breath as more smoke billowed up about him. Holding his hand over his face he crossed to the main current breaker on the wall and pulled it to 'off', but was forced to retreat by the increasing volume of smoke pouring out of the top of the oven.

Putting his head outside he sucked in another lungful of air and with eyes streaming, groped his way over to the water tap on the preparation basin. He found it and wrenched it on, but nothing came out! The water pipe into the building had been shut off. Desperately he ran out and across the deserted campus to his house.

'Fire! Fire in the plant presses!' he shouted at Anne and Eduardo Lleras, who were just finishing their drinks on the patio, and then made for the kitchen where, scattering dirty dishes left, right and centre, he grabbed a saucepan and started to fill it with water. Eduardo came in and, taking another pan, did the same. Anne rushed off to raise the alarm further afield.

Eduardo and Iain staggered down to the botany building, a hundred metres or so across the campus, carrying slopping saucepans to throw water on the fuming oven. They ran back for more. After two more trips staff and students, roused by Anne's calls, ran up to help them. One of them flung open the oven door to see if he could save some of the plants, but the inrush of air caused the fire to flash round and red flames now shot up brightly from out of the burning oven. Back and forth went the relay of saucepan handlers, tiring more and more at each run.

Meanwhile, the INPA resident wood technologist, hearing the shouts of alarm, rushed into his own kitchen and grabbed the largest saucepan he could see cooking on the hob. Hastening down to the oven he flung the contents onto the flames, by now licking at the ceiling. A sticky white mess slapped onto the side of the oven and slid down to the floor. He had just thrown his supper – a pot full of tapioca – onto the fire!

At the same time one of the course students had the idea of using his car fire extinguisher on the flames. But he didn't know how to set it off – he pulled rings and hit knobs but nothing seemed to work. He looked at it more closely – and it went off in his face. He shrieked and was rapidly covered by a layer of thin white foam – making him appear ghostly in the darkness. Just at this moment his wife, who had joined the growing, excited crowd of spectators, spotted his features unearthly pale him looming out of the darkness. Being something of a superstitious woman she screamed and started shouting 'Pedro's dead! Pedro's dead!' adding greatly to the confusion.

Throughout, Iain and Eduardo continued to trek back and forth with their saucepans, now grunting like athletes at the end of a marathon. Fortunately, at last someone ran out a hose and the steady stream of water dealt a death-blow to the flames and in turn resurrected Pedro, much to the relief of his wife.

Iain and Eduardo staggered back to the kitchen for the last time and dropped their saucepans into the sink and themselves into chairs. 'I think,' said Eduardo, 'that if I have to do that just once more I too will be dead!' Iain, panting and covered in sweat from the heat and the exercise, heartily agreed, but, as ever, having caught his breath went back down to the preparation room to survey the damage – which was not as bad as they had feared. The room itself had not suffered structurally (though the ceiling needed a good clean and paint) and oddly, neither had most of the collected plants inside the presses (the most valuable consideration by far), though the oven itself was a smoke-blackened mess.

'INPA was such a wonderful place to practise tropical botany,' says Iain. 'It still is. Whenever you are in the Amazon you just never know what is going to happen next!'

17. BITING THE BIG APPLE

Once Iain had seen his first class of Amazonian degree students successfully through INPA and inducted the second, in many ways he would have liked to stay in Manaus and continued there as course director, if it could have been approved and arranged. But he discovered pretty sharply that the New York Botanical Garden intended reclaiming its own, with something of a vengeance. The president, Howard Irwin, wrote to offer him the post of director of scientific research in New York.

As with many others who have been in a position to influence Iain's career down the years, Dr Irwin had watched at a distance while Iain quietly got on with something far removed from most people's attention – and produced good results. This showed not only that he could manage but also that he could develop good ideas and follow them through, and more important than even this: that he was not self-serving or ambitious (except for the advancement of science) and so had the capacity to take on responsibility in a cool and undistracted manner. He would not prove a threat to his superiors, or hold back his juniors.

Dr Irwin's invitation was a wholly unexpected and quite remarkable offer to climb quickly up several rungs of the promotion ladder. Until now Iain had been a research assistant and then (the newly created) B.A. Krukoff Curator of Amazonian Botany, and, though during his time at INPA this had been suspended, he had every reason to expect to drop back into that position once more. But now a quite different order of work was on offer – a chance to lead the whole of the botanical science research programme at the NYBG.

The research director had the responsibility for devising and executing all the botanical outreach of the gardens – and that was considerable: over seventy scientists and numerous research projects at home and abroad. More important even than that he held the whip hand in terms of research policy. If the research director woke up one morning and felt that investigating tundra in the Arctic Circle was wholly significant, then in fairly short order a good number of NYBG botanists would be fitting for snow-shoes. He might well have to justify it, but the basic principle was that what the research director directed, happened. It was a position of powerful significance in the US scientific community and the botanical world at large.

But though honoured and excited at this unexpected opportunity, Iain did not immediately feel that he should accept it. His concerns and heart were in the Amazon and although much of the work he could expect to direct would have to do with South America, would the job permit him to have any hands-on work there? Or would he be expected to be tied to a desk in New York? For Iain this was crucial. To have the chance to influence the direction of research at NYBG was one thing – something Iain would greatly value – but at the loss of his beloved rainforest? Some things are not worth sacrificing. So on his return he discussed matters in some detail with Howard Irwin, who, certain Iain was the man for the research director's job, was able to offer him what he wanted. Yes, he would have to spend more time in New York, but he could still go out in the field from time to time on Amazon expeditions – just as long as things ran on smoothly back home. He could certainly do front-line research – provided that he did not neglect the oversight of others or the crucial business of raising funds for the NYBG research projects.

Although in many ways this was tantamount to saying he could do what he wanted as long as he did the traditional research director's job first, it was enough for Iain and on this basis he accepted the post.

So Iain, Anne and the children set up home – at last – in their new house in suburban White Plains. Iain was now directing the work of all the plant scientists in the botanical garden of the prime city of the United States eastern seaboard.

He quickly drew a team of supporters round him – to do all that he wanted (personally and as research director) he was going to need a lot of help. Head of the list was his secretary for five years, Mickey Maroncelli – a vibrant New Yorker from Yonkers who, as a Roman Catholic, had a tendency to trade Bible texts with Iain on whatever subject was under discussion. Not that she took his faith lightly – she recalls that she always knew when a big decision was pending for her boss: the inner door to

Iain's office would be gently closed and his Bible would be taken from the top drawer of his desk for a few minutes of silent reading and prayer.

'It wasn't for long,' she says, 'but then he seemed to discover a sense of peace and rightness about what was to be done in a situation. It didn't matter what it was: it could be as serious as asking someone to leave because they couldn't do their job – something he hated to do but never shied away from – or knowing what to say in an important symposium. He was never other-worldly about it; just concerned to do the right thing by God.'

She also remembers his remarkable lack of dress sense: 'His mind was on getting the right plants, not the right pair of pants!' He would often come in to work with mismatched socks and one day even arrived wearing a boot on one foot and a shoe on the other. She only spotted this when she saw him limping unevenly along the corridor towards the herbarium! His defence was that as he had got up early to come into the office he had tried to leave home quietly so as not to disturb Anne and the children. This meant fishing around in the wardrobe in darkness and putting on whatever he could find. She didn't believe a word of it – and Dr Scott Mori, another member of the team (now senior curator at NYBG), remembers travelling with Iain in the field and finding him similarly preoccupied with plants at the expense of his own personal comfort.

'On one expedition he slept in a rotten hammock which slowly gave way throughout the night, one rope at a time. All the rest of us could hear his hammock lines going "ping" every half-hour or so, but he slept through it all – and woke up bright and early the next morning about two inches from the ground and bent nearly double!

'He was once so keen to go out collecting that he ate his breakfast oatmeal porridge dry, without milk or water, covering his famous beard with white flecks, like wedding confetti. I promised to buy him a horse's nosebag when we got back to the States, so he could start out collecting without stopping for breakfast. He could eat going along!'

With his personal research, a whole department to organize and expeditions to mount, tension in Iain's office sometimes mounted to high levels, especially when deadlines for research papers or Science Foundation applications approached. Then Mickey would resort to 'defusing' tactics – a favourite was to fire Iain's overseas correspondence onto his desk by making paper darts out of it. 'That made sure it went airmail!' she says.

Once she turned up (on 1 April) in the office dressed as a nun. This, she told a rather surprised Iain, was because her husband was still recovering from a back injury after many weeks and had therefore been unable to perform his full marital duties for a long time. At least her church gave her a traditional way out in such cases!

Despite the banter, both she and her husband deeply valued Iain and Anne's personal Christian concern when she lost a close relative – both of them came independently to see her at home and bring flowers and comfort. They also attended her daughter's wedding. She has never forgotten their love and support on these occasions.

Mickey, and her assistant in later years, Rosemary Lawlor, had to take much of the responsibility for the smooth running of the department, for, as he had agreed when taking on the research directorship (and determined to do), Iain went south at least once every year for his 'fix' (as Anne called it) of Amazonia. In fact a number of important expeditions followed this promotion and throughout the following ten years Iain spent quite a lot of his time in the field, though a lot of this time would be spent sending back memos and letters, reports and suggestions on research policy or departmental matters which Mickey and Rosemary translated into typed advice and instructions to be sent round the department.

Although Iain had moved into the executive arm of the garden's organization he still made a point of identifying with the scientists, and kept his office in the science department rather than moving to a smart executive suite on the administration floor. He spent as much time as he could in the herbarium and in hands-on science himself and he joined the traditional, informal 'sandwich lunches' in the herbarium foyer, so that by chatting with his teams he could keep informally abreast of their work and, more importantly, sense the mood of the department.

He is remembered in NYBG as 'the man who liked to say yes' – ever keen to encourage new ideas, although this enthusiasm to involve others left him very short of time himself. New post-doc students interested in working under him were sometimes told that they would certainly learn a great deal if they could get hold of him. The story is told of one post-doc student who decided that the only way she could get her thesis discussed was to stand around outside the men's washroom – knowing Iain would have to show up there sometime! Nevertheless he helped eleven students to gain their PhDs during his time in New York.

Brian Boom – another colleague (now NYBG vice-president for botanical sciences – Iain's last post in New York before coming to Kew) recalls Iain's legendary capacity for work:

'He was a workaholic. He would arrive at the office or herbarium at six-thirty or seven in the morning and work through until late at night, sometimes into the small hours. He always said – rightly – that he never asked more of others than he did of himself, though knowing Iain he wouldn't have found that very difficult. To me one of his greatest strengths was that, whatever the pressure or circumstances, he always seemed to find the right thing to say.'

A major thrust of Iain's work, commenced soon after his return to the NYBG in 1975, was 'Projeto Flora Amazonica' (later 'Programa Flora Amazonica') – an idea of Iain's to make a US–Brazilian bi-national project out of a Brazilian initiative intended to produce a computerized database of the plants of the Amazon region. They planned to put all the information from their national herbaria onto computer, train young Brazilian botanists and compile further tropical plant monographs.

Projeto Flora Amazonica was a great project in every sense, not only advancing botany in Brazil, but firmly cementing the connection between New York, Brasília and Manaus, which Iain had long wanted. In terms of work over the ten years of its operation (1975–1985), 300,000 collections were made in new regions of the rainforest (which are still being worked on in herbaria today), and over 1,000 new species were described – with a generation of fresh new scientists from both North and South America receiving training in practical as well as theoretical tropical botany.

The project produced many specimens and gave much extra work to the director of the herbarium, Dr Pat Holmgren (the wife of Noel Holmgren, the former student who had nearly lost Iain in the forest on his first field day in Surinam, and now working at the NYBG). Pat's organizational abilities and attention to detail exceeded even Iain's and she worked closely with him on Projeto Flora (despite being a specialist in temperate botany) often deputizing for him whilst he was in the field. He readily acknowledges that without Pat he would have achieved far less, both in New York and in the field.

Another old face in New York, and a solid supporter of Projeto Flora, was Enrique Forero, a Colombian who had been Iain's first postgraduate student back in 1971, though he had met him on expedition as far back as 1967. Then Enrique had only been able to speak Spanish – and Iain only English – but Iain says that 'from the start we got along fine'. So much so that Iain put him in charge of the land team on his expedition up the Madeira River in 1968.

Iain took on and organized the NYBG's US bicentennial commemoration symposium in 1977. In keeping with his ever-growing concern for world ecology he suggested the theme 'Endangered Plant Species of the Americas (From Canada to Argentina)'. In the celebratory atmosphere of the bicentennial many thought that this might be too downbeat. But Iain felt he was voicing his, and others, heartfelt concerns. The time was ripe to say something and say it loudly, not just for the rainforest, but for all of suffering nature. Fortunately Howard Irwin agreed enthusiastically, as he was equally concerned and authorized the symposium – and in the event it proved to be the genesis of much serious thinking in the American botanical community.

'Extinction is Forever' became the watchword of the international gathering and was used as the title of the edited proceedings which Iain and his colleague Tom Elias (now president of the US National Arboretum in Washington) put together afterwards. Far from proving the 'downer' many had feared, all who attended appreciated the sober realism of Iain's approach, set as it was against a background of the historic struggle for survival and independence in the former American colonies. Many saw that once again their heritage was in need of active defence. This time though it was the natural heritage: the plants and trees of their breath-takingly beautiful continent. 'If it was worth fighting for then, it is worth fighting for now,' was the concluding mood of the delegates, all of whom left New York with a reaffirmed resolve to prevent as far as they possibly could the extinction of species the symposium had highlighted.

Another milestone of Iain's time in New York was the 'Way Ahead for Botanical Gardens' conference called by Jim Hester when he took over as NYBG President in 1980. At the time, Dr Hester had a limited understanding of botany, though considerable experience as an administrator of university faculties (he had previously been president of New York University and rector of the United Nations University in Tokyo), and wanted to identify new issues facing botany as well as learning more about the subject – as the experts discussed and debated.

He posed a simple question to his scientists, heads of department, horticulturalists and managers: 'What ought we to be doing here at NYBG that we are not?'

From the discussions it soon became evident that there were three possible new directions they should be considering. Firstly, the increase in environmental awareness among the general public had forced scientists of many disciplines to look to their laurels on matters ecological. Iain, now much taken with ethnobotany, felt that the whole relationship between people and plants in economic terms deserved a closer look. This was championed by Professor Richard Schultes, director of the botanical museum of Harvard University. And finally there was botany within cities – the profound difference plants made in the lives of city dwellers was not in doubt, with half a million visitors to the New York Botanical Garden alone every year. But who knew about city plant life? What really happened to plants in compacted soil, in traffic fumes, in window-boxes and high-rise apartments? It too should be studied.

Iain was asked to look into the setting up of an Institute of Economic Botany – to study plants in the service of humankind. Initially an odd mix of disciplines (to the purist) of geography, commerce, anthropology, ethnobotany and agriculture, this was to become a specialized science at NYBG. If, of course, someone could be found to fund it. For in a world increasingly aware of damage to forests, of desperate droughts and famines (this was to be the decade of Band Aid), of the divide between rich and poor, north and south, understanding the world's economic uses of plants (trees, crops, medicines, and so on) was now vital; for humankind *and* the plants.

Also set in motion was a project which became the Institute of Ecosystem Studies, based at the NYBG's Cary Arboretum, as it was similarly felt that eco-systems needed attention in their own right. Lastly a pilot plan was formed to look into plants in the urban environment.

Iain had to put on his best persuasive sales manner and, in the same way he had done for his first sponsored trip to Turkey as a student, he and Dr Hester started approaching various institutions for the many hundreds of thousands of dollars of grant money needed to fund his new institute. They finally came upon the Mellon Foundation who agreed to do just this and so Iain found himself with yet another team to direct.

Over the years the Institute of Economic Botany has grown greatly and is now headed by Iain's one-time assistant there Mike Balick. From a staff built up to ten when he left the NYBG in 1985 to twenty-four today. Its work has had much that is practical, as well as theoretical, to offer to both scientific and general understanding of the value of plants to people. Many of the projects to utilize the resources of the rainforest in a constructive (instead of destructive) way have stemmed from research conducted by the Institute. A classic is the babassu palm (*Orbignya phalerata*) project which concentrates on growing these palms in parts of north-east Brazil so that their nut husk can be used as a renewable fuel resource (among other things), thus avoiding the need to fell so much of the (non-renewable) rainforest to produce firewood and charcoal fuel. The babassu also produces an edible oil and a flour. It is a wholly effective illustration of the value of the relatively new science of economic botany.

A precursor to this was an interesting piece of botanical history which Iain had researched. He took a month or two from his herbarium work to produce what became one of the most useful tools for Amazon field botany for a number of years – particularly so for Projeto Flora Amazonica.

Looking at old documents and maps, Iain drew up a historic itinerary of all the routes taken by Amazon botanical explorers down the ages. He included all the recorded routes of intensive nineteenth-century investigators such as Richard Spruce and Von Martius back to the earliest Portuguese explorer Alexandre Rodrigues Ferreira (1756–1815). This became a paper called 'An Index of Plant Collectors in the Brazilian Amazon', published in the very first issue of *Acta Amazonica* in 1972.

Now planning expeditions could avoid these routes and so fill in the botanical map more thoroughly. The information proved so valuable in Brazil that others started doing the same for their own countries notably in Peru and Venezuela.

Though Iain did still travel, by and large the family now stayed in New York, as unbroken schooling was increasingly important for the girls. (Rachel was sixteen and Sarah eleven.) Both discovered that they had been served well by the Brazilian school they had been attending in Manaus. They proved to be ahead of their year in science and maths subjects when they went to school in the United States.

Anne, relieved of the need to run courses or expeditions chose to take up work as a personal tutor of English, when she wasn't caring for the girls at home. She taught expatriate (non-North American) visitors

Iain and Scott Mori examine a fruit of *Lecythis* in the rainforest of French Guiana. This is in the Brazil nut family which they studied together

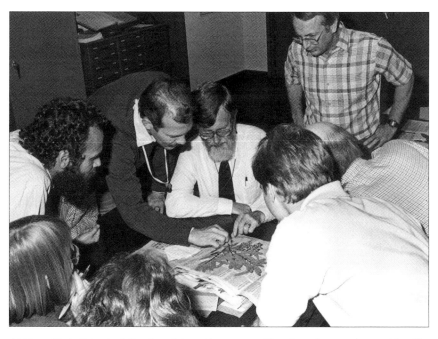

At New York Botanical Garden: Iain with some staff and students trying to identify the family of a collection

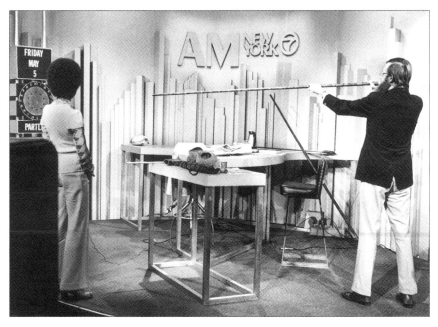

Demonstrating an Indian blow-pipe on a New York chat show

Ian usually drove one of the expedition boats himself

A *Rhabdodendrum macrophyllum* bush, a genus for which Iain described a new plant family, Rhabdodendraceae

Her Majesty Queen Elizabeth the Queen Mother opening the newly restored Palm House at Kew in 1990

HRH the Prince of Wales visiting the Jodrell Laboratory in 1993 and having a demonstration of molecular systematics from Mark Chase (right) watched by Michael Bennett (left)

Professor Gren Lucas and Iain explain the biology of the *Victoria* water lily to Prime Minister Margaret Thatcher during her 1989 visit to Kew

Sir David Attenborough opened the new Seeboard Study Centre at Wakehurst Place and then fascinated the schoolchildren by taking them for a walk in the Loder Valley Reserve

Opening the new Kew shop
in 1992 with Anita Roddick,
founder of the Body Shop
and keen conservationist

Spreading the word – Iain
lecturing on conservation at
Kew, 1993

Sir Robert and Lady Sainsbury receive the first Kew Award in 1994 in recognition of their support of the orchid work at Kew. Left: Robin Herbert, Chairman of Kew's Board of Trustees

Iain with Armon Takhtajan when he opened the restored Wakehurst Place Arboretum which is arranged after the Takhtajan system of classification of biogeographic regions

The newly restored Chokusi-Mon (Gateway of the Imperial Messenger) and a new Japanese landscape at Kew was opened by HIH Princess Sayako in the presence of HRH Princess Alexandra in October 1996

Iain welcoming HRH The Prince of Wales who attended the launch of the Millennium Seed Bank appeal in 1996

Iain and the Rt. Hon. Michael Heseltine, Chairman of the Millennium Commission, perform the topping out ceremony of the Millennium Seed Bank Building at Wakehurst Place

to New York. A large number of governments and international corporations appointed representatives to New York and many settled their families in White Plains. Anne had never forgotten her early days in America and sought to help her students not only to master the language, but to adjust to the American culture. An outsider herself, she seemed to know just what was on their minds as they struggled to cope with their new way of life. Her waiting list for tuition was three times over-subscribed. The Prances too were away from their home country, however successfully they had adjusted to America, and they did what they could to keep up with relatives in England – visiting them on whistle-stop tours roughly every three years. Sarah Prance recalls that these were odd holidays: 'We slept in a different bed every other night and spent the daytime repeating the same things to different people!'

Not that she minded. Rachel and Sarah feel that they had a very stable and happy upbringing, which may seem surprising given their many moves. They felt that their parents gave them as much time as they needed, and, what perhaps was more important, that the time when given was wholly theirs.

Their father was often away, they were used to that, but when he was home he made sure they had a good time. He read bedtime stories to them (some made up by his mother) or invented them himself, and he would sing to his children some of the Gaelic songs he remembered his mother singing to him, except that he would change the words to introduce characters from the girls' bedroom – favourite cuddly toys or dolls – and make the song all about them, much to their delight.

At their birthday parties he would invent children's games far superior to 'pass the parcel' or 'musical chairs'. Iain would have his daughters and their friends posting things in different letter boxes, carefully numbered around the house, or deciphering complex clues for buried treasure in the strangest of places. But his *pièce de résistance*, the thing that made a Prance party the very best on the block, was the Amerindian blow-gun game.

Iain would get down his blow-guns from the Amazon and line the children up behind a row of chairs in the front room. Then across the room (or garden in good weather) he would place a line of balloons. The children were given a blow-gun, which was so long it rested on the chair backs, and a dart. They would fire away, sending their darts singing through the air to pop a balloon, if their aim was good. This was real Amazonian fun and quite unbeatable. The trouble was that at the end of the party all the adults who came round to collect their children wanted to join in, which rather stretched the supply of balloons. In fact blowing blow darts at balloons (or sometimes pumpkins) became something of a Prance party motif in those days and many was the couple invited over to dinner who ended the evening in the backyard puffing for all they were worth trying to explode a balloon with an authentic example of Amerindian economic botany.

The Prances' house looked rather like the atmosphere Iain and Anne tried to create inside: American, but with a strongly British flavour. In an open-plan neighbourhood of weather-boarded wood houses the Prance home stood out by being compactly built of red brick with a neat chimney-pot to serve a real fire below in the cosy living-room. In the backyard Anne planted and kept an English cottage-style garden with fresh herbs, which she always used for cooking, sown in among the flowering plants. The Prances would always make wherever they lived, be it America, Brazil or elsewhere, very much their own.

Recently when Anne visited her daughter Rachel, now working on a project for Aids victims in Recife, Brazil, she noticed the neat, characterful home Rachel had managed to make out of the very small apartment she was renting in the poor quarter of the city where her work was.

They continued to attend the White Plains First Baptist church and took a full part in its life, though often as a family their international way of thinking caused a few raised eyebrows. Once, when they felt that the personal antipathy between Russians and Americans had reached ridiculous heights with the escalation of the cold war, they involved their church in a city-wide goodwill project to make half a million paper birds to send as a peace gift to the children of Moscow. These were to represent prayers from the children of one nation to the children of another, and their mutual desire for peace.

The idea came from a story told after the Second World War when a Japanese child in Hiroshima, suffering from radiation sickness, started to fold origami paper cranes. Legend had it that to create a thousand of these paper birds, highly regarded in Japan, meant life for the creator. Sadly the little girl died before she could finish her task but her idea grew to represent every child's will to live in an age of potential nuclear disaster.

Allied to the making of the paper cranes were discussion seminars, Bible studies and information packs on the value of co-operation as well as confrontation. The climax of all this was a service of dedication in a large church in downtown Manhattan (which had originated the project) where teenage Sarah Prance, today a medical doctor in London, helped thread the thousands of cranes onto string and wire and hang them criss-crossing the church, and up into the far heights of the roof.

It was a gesture, no more, but because of it she feels many of her friends and neighbours were moved to think more about the common humanity they shared with others, rather than just their differences, as they looked across the great divide that was the Iron Curtain. In conservative New York suburbia such things were certainly food for thought.

18. A GARDEN IN THE CITY

The New York Botanical Garden, Iain's scientific home for nearly a quarter of a century, was established in 1891 by a group of philanthropic New Yorkers under the guiding hand of Nathanial Lord Britton, a visionary botanist from Columbia University. Following a visit to Kew Gardens in England, on honeymoon with his botanist wife, he returned to New York determined to set up an institution there for a similar purpose. Others on the newly-formed Board of Managers included such notable figures in New York society as Cornelius Vanderbilt and Andrew Carnegie – so from the start the projected botanical gardens had the firm support of powerful local patrons.

Their early vision was to preserve plants, present botany and offer a restful garden to the burgeoning city. The 1891 State Charter stated:

'A body corporate... for the purpose of establishing and maintaining a botanical garden and museum and arboretum therein... for the collection and culture of plants, flowers, shrubs, trees and the advancement of botanical science and knowledge and the prosecution of original researches therein and in kindred subjects, for affording instruction in the same, for the prosecution and exhibition of ornamental and decorative horticulture and gardening, and for the entertainment, recreation and instruction of the people.'

All of which, while not perhaps being quite as snappy as the modern NYBG mission statement, still covers pretty well the scope of their present, as well as their original, activity and purpose.

The board took over land straddling the Bronx river which had belonged to a prestigious family who had made their money, as so many did in colonial days, out of tobacco. The Lorillard family milled tobacco snuff in Manhattan since the 1760s and moved to the banks of the Bronx river – a three-metre wide stream well capable of driving their water mill – in the 1840s.

They sold some of their estate to New York City in 1884, having moved to yet larger premises in New Jersey. This acreage became the first New York Botanical Garden in 1891, with the original Lorillard snuff mill, carriage house and stone cottage being included in 1938 to form the present 250-acre site. The grand Lorillard mansion had burned down in 1923.

The old Lorillard snuff mill now holds pride of place in the gardens down by the river, restored and declared a US national landmark. It is a handsome if clearly utilitarian building, used as a garden restaurant, sitting demurely on the banks of the Bronx amongst the trees of the forty-acre Hemlock Forest – the only remaining uncut woodland in New York City.

These peaceful surroundings are a far cry from the old days of active commerce at the mill, when bale upon bale of tobacco leaves would be hauled in for milling, piled high on the teamsters' great horse-drawn wagons which rattled and clinked along the dusty trails through the hemlock woods. The tobacco would come from upstate plantations or from the New York waterfront, landed by the sailing coasters which plied the eastern seaboard. And each bale would be quickly opened and carefully inspected by Paul Lorillard before being unloaded, all to the background noise of plash and plunge from the water-wheel, the shouts of the mill boys and the gruff commands of the master miller, the continuous grind of the millstones and a general coughing and sneezing from both men and horses in the spicy, snuff-laden air.

It had been a busy, relentless place. But the bustle of commerce is now kept well away from this quiet green oasis of calm, except that now and then the old and new do combine: the T.A. Havemeyer Lilac Collection for example, just across from the preserved stone cottage – part of the original Lorillard commercial property – surrounds the first site of the Lorillard 'Acre of Roses'. These were grown not for horticultural enjoyment or their romantic appearance, but for their unmistakable scent which, as ground-up rose petals, was an essential ingredient in the snuff produced at the mill. The gardens' main collection of roses no longer grows in this location but immediately next door, in the Peggy Rockefeller Rose Garden.

This was originally designed by Beatrix Farrand – a noted early twentieth-century garden designer much in demand with the leading families of the city – and begun in 1916. Following its removal in 1965, it has now been fully restored and, surrounded by the original wrought iron fence, once again wafts to the passer-by the distinctive scents that Paul Lorillard so much needed for his snuff. It also offers the wonderful sight of many varieties of possibly the most-loved and talked about flower in the world.

Never in their wildest dreams could the nineteenth-century Lorillard gardeners have imagined, as they mulched, pruned and toiled over their master's thorny 'product ingredients' that one day their work would be recalled and recreated in a riot of colour by some of the most highly gifted gardeners in the land.

The Bronx river itself, the factory's essential energy supply, now originates near the Kensico Dam in Westchester County and joins the East River near Hunts Point in the Bronx. Just upstream of the mill, within the gardens, there stands a two-metre dam and fall, built to stabilize and regulate the thrust of the water stream which powered the great wooden water-wheel.

Next to the snuff mill and taking pride of place in the botanical, as opposed to architectural or commercial, history of the gardens, lies the forty-acre virgin Hemlock Forest, preserved solely through the existence of the NYBG as the last remaining representation of the arboreal flora that once grew across the area now occupied by New York City. Trees there include the classic mix of american beech (*Fagus grandifolia*), red oak (*Quercus rubra*), cherry birch (*Betula lenta*), tulip tree (*Liriodendron tulipifera*), white ash (*Fraxinus americana*) and, of course, hemlock (*Tsuga canadensis*) which once crowded down to the shores of the Hudson river and thickly covered Manhattan Island.

Described in the early years of the Botanical Garden as 'the most precious natural possession in the City of New York' the Hemlock Forest is still actively used by the Science Department, as well as students and visiting botanists, for a number of arboreal projects, mainly relating to the trees' ability to survive and adapt in the surrounding urban environment. Not that from there the city is much in evidence. To stand on the pretty arched High Bridge and look down along the river as it gently winds its way into dappled sunlight past the trees of the forest is definitely an experience of New England countryside, not a city park.

Surrounding the Hemlock Forest which lies in the centre of the gardens is the perimeter road. It starts as Magnolia Road and continues as Snuff Mill Road and Daffodil Hill, curves through many specialized areas of trees and plants – magnolias, legume trees, beech and oak trees, dogwoods – on past the Children's Garden into Cherry (tree) Valley and the Havemeyer Lilacs and Rockefeller Rose Garden to the snuff mill and the river. Here not only plants catch the eye, but sharp outcrops of rock – strange, striated masses, intruding into the otherwise gentle landscape. And here and there stands a lone boulder, left by the Wisconsin glacier which receded some 15,000 years ago.

Snuff Mill Road crosses the river and branches many ways, through the forest, along the riverbank – which leads out to the Bronx Zoo next door – or up Daffodil Hill Road – a cascading flood of yellow and white in springtime – passing the unique collection of ornamental crabapple (*Malus*) trees and the groves of white pines.

The white pine tree (*Pinus strobus*) was a common sight to the early colonial settlers in North America as it grew in vast numbers in forests right across the north-east. It was much valued in construction as it is a straight-grained softwood, making it easy to work, and has a pleasing light colour. With its distinctive long pine needles and pine cones it is a real former 'local resident' in the gardens. White pine is still used extensively for timber in the eastern USA, though the great forests have long since disappeared.

No discussion of trees at the New York Botanical Garden would be complete without including a mention of the Cary Arboretum – a wide 2,000-acre tract of land well away from New York City, in Millbrook, Duchess County, New York State.

This was acquired in 1971, being awarded to the gardens by the Mary Flagler Cary Charitable Trust which then owned the land. At the time the NYBG was very keen to find a site to develop ecological and environmental sciences. The existing Bronx garden site afforded few opportunities for complex or large-scale natural ecology experiments to take place. First because it is a city site and has many visitors, and secondly because, by definition, it is a deliberately cultivated garden. For a truly normal and natural environment they needed to look further afield. The Cary Arboretum fitted the bill perfectly. Now an independent research institute, called more formally the Cary Institute for Ecosystem Studies, research projects undertaken here include studies of tree diseases, including the breeding of disease-resistant trees (particularly of American elm) and projects to breed trees better able to resist urban pollution. All ecological interaction is of interest too, particularly the ways insects and plants work together. There are practical projects too to study the resident animals, such as the feeding habits of young deer, so as to learn how to protect valuable park plants and trees from unnecessary animal damage.

The Cary Arboretum has also pioneered ways of constructing environmentally-friendly buildings. The Plant Science Building, completed in 1978, was the vision of Howard Irwin and Robert Goodland, and was one of the first purpose-built buildings to employ natural insulation, heat recycling systems and extensive solar panelling.

Back in the Bronx, the Daffodil Hill Road curves past one of the three garden lakes and rises to meet Conservatory Drive opposite the Kennedy (Main) Gate. Just here stands the most striking and apparently 'botanical' building in the whole garden complex: the gleaming glass and iron 'Crystal Palace' of the Enid A. Haupt Conservatory. This was constructed between 1899 and 1902 and inspired by the Victorian Crystal Palace in London, though built to do the work of the famous Palm House at Kew, which it partially resembles.

It has a thirty-metre high round 'Palm Dome' in the centre and two lower adjacent wings for Tropical Flora, the Fern Forest and seasonal displays. These side wings turn at right angles to run south-west and then at right angles again, towards each other, enclosing a large rectangular courtyard. These extra wing sections, with elegant squared domes at each of the corners, provide additional glass cover for the Hanging and Containerized Plants section, the Orangerie (citrus fruit displays), the Economic Plant area and Green School – for ecological and educational exhibits. Opposite these lie the Old and New World Desert areas and the Special Collection (Australasian and South African plants, among others).

In the courtyard enclosed by the estimated 17,000 panes of shimmering conservatory glass lie both a tropical and a temperate pool for various aquatic plant exhibits, including the Chinese sacred lotus (*Nelumbo*), water lilies (*Nymphaea*) and of course *Victoria amazonica*.

Iain recalls many meetings with Enid A. Haupt herself – local city philanthropist and long-time member of the New York Botanical Garden board of managers. She represented to him all that was good in a very long tradition in the United States of personal patronage for the arts and science. An heiress from the Annenberg family, who made their fortune as railroad entrepreneurs, her firm conviction was simply that plants helped people. Therefore, she reasoned, people should be given as much opportunity as possible to be with plants, in schools, in hospitals (to which she also extended much of her patronage) and above all in the New York Botanical Garden – which is always open to the public.

She was a 'stunningly elegant' lady, Iain recalls, although very much the senior citizen by the time he came to know her.

'She always dressed to perfection – even the tint of her spectacle lenses would be carefully chosen to match her co-ordinated hat, dress, shoes and handbag!'

A woman so cultured, perhaps, that she was the exception that proved Iain's unstated rule that the more sophisticated someone became the less they were likely to think about plants and the earth. Certainly, with her active interest in and support of her city's major botanical institution, no one could accuse her of anything of the sort.

The gardens of the New York Botanical Garden form even more of an oasis of beauty and peace in the midst of the city than does Kew in London, for in New York the Bronx area, as most people know, represents much of the grimmer side of inner city life. The NYBG managers take great pains to encourage local people to experience and enjoy the gardens, with a low gate tariff. 'Pay as you wish but you must pay something' is the rule (the City of New York covers only about 20 per cent of the gardens' running costs), and a directly accessible 'interpretation programme' of explanatory signs is supplied, as well as garden directions and a patrolling, uniformed security service to advise visitors and keep the gardens safe.

To go from the dour, forbidding streets of the Bronx through the wide main gate of the Botanical Garden (off the Moshulu Parkway) and see a swathe of tended green sweeping up between the lines of trees to the imposing NYBG Museum Building is a moment of instant relief and recreation. The Museum Building is the headquarters of the gardens, containing the visitors' centre, with its stunning display of orchids under a glass rotunda in the foyer, herbarium (with over five million specimens), research offices where Iain worked, the large botanical library, and extensive garden shop.

Behind this, a monument to a less stylish period of architecture, the sixties monolithic block of the Watson Building houses the administration, members offices, and the Institute of Economic Botany which Iain founded and where many of his botanical colleagues were – and many still are – based.

Science and education are a success story at the NYBG which probably leads the world with its programme of general botanical education. Most botanical gardens (the NYBG is among the top five) have various graduate programmes for students in botany, horticulture or related disciplines but New York have widened their educational programme to include people from all backgrounds and academic levels.

Students may enroll in evening or spare-time classes ranging from mycology (the study of fungi) to flower cultivation, molecular biology or landscaping and ecology. Bruce Riggs, Resident Horticulturist (and incidentally an Iain Prance look-alike, which has caused more than one case of mistaken identity) notes that over 14,000 adults and 53,000 children join these programmes every year, quite apart from those who visit the gardens as part of other educational initiatives.

Although mostly kept away in the herbarium or science offices in the Watson Building, or visiting scientists in the nearby Harding Laboratory or the Propagation Range by the snuff mill, Iain's love of living plants often drove him out to the grassy spaces, groves of trees and rockeries of the wider gardens, to lead tours for special visitors, or to walk and think and thank God for all the green planted there by the early visionaries to relieve the city grey.

As Gregory Long, current president of the New York Botanical Garden says:

'People do *need* plants. In our own way we are as vital to New York as Wall Street. Not only because we are here *in* the city, but because of what we do – in education, in science and in exploration. For twenty-four years Iain Prance was a highly valued contributor to this work. He is an eminent botanist, and a good friend and advisor. The city – and nation – certainly owe him a great deal for all the work he did here. Many wish he could have stayed longer.'

In the spring of 1987 leaving the New York Botanical Garden could not have been further from Iain's mind. Created a vice-president of the Garden in 1977, while still retaining his post as director of botanical research, he was later promoted to senior vice-president for science in 1981, continuing to split his time between his many duties in New York and his tropical interests in the Amazon.

The large-scale Projeto Flora Amazonica had just drawn to a close and he was actively seeking funding for spin-off projects, florulas, to study certain areas of the Amazon rainforest more closely. The Institute of Economic Botany was in full swing proving a solid and effective springboard for much new botanical science directly affecting and helping people.

Then one day a letter arrived. On the envelope was emblazoned the crown, lion and unicorn of the British royal coat of arms. It was postmarked Kew, England.

19. ROYAL APPOINTMENT

The arrival of the letter from the Board of Trustees was not wholly unexpected. But Iain was nevertheless surprised at the offer it contained. Some months before he had received copies of an advertisement for the post of director from no fewer than six colleagues! So, taking the hint, he had applied for the position and been shortlisted with three others. But, following the final interviews, he had heard nothing more.

Thinking he had not been successful he dismissed the matter from his mind, until the distinctively postmarked envelope plopped onto his doormat one morning in White Plains. It asked, quite simply, if Iain would consider taking up the post as director. The delay had been due to the letter coming by sea, not airmail. (By this time back in England the Kew board were getting worried that their first choice of candidate was having second thoughts!) They all felt that in Iain they had made an excellent choice.

Iain's reaction was typical. While highly gratified, and not a little amazed, that he should have been accepted for so elevated a botanical position he did not rush either to refuse or accept. He took time to consider, though out of courtesy he immediately informed the president of the New York Botanical Garden of the offer and outlined some of the considerations that might make him inclined to accept it.

What he was not prepared for was the reaction in New York. All of a sudden it seemed as if they realized what they were about to lose if Iain went back to England – not only a significant scientist and vice-president of the NYBG but a major asset in Amazonian botany; a world-class expert in one of the most significant regions of the world. And this particularly at a time of renewed public interest in the Amazon, the deep commitment of the NYBG to South American botany in general and the undoubted co-operative success of projects such as Projeto Flora Amazonica.

Hurried meetings were held and Iain was called to discuss various proposals over lunch in graciously appointed Manhattan apartments. Was he perhaps looking for some more responsibility in the NYBG? Did he have any projects he felt needed more attention than they had recently been getting? Was his salary sufficient? The NYBG would certainly give him a raise, a substantial one, if he would stay on. There was even the possibility of him taking over as president. The current NYBG president might well be happy to consider taking retirement a little earlier than he had intended. Then Iain could gracefully step into his shoes.

If Iain was ever a man to respond to attention and flattery then this would have been the time to do so. But Iain was not such a man. As Anne Prance says:

'He never plays all the little games other people generally expect or enjoy. It would never occur to him. He just thinks about things, plans and then gets on and does them. Curious really!'

In fact Iain was mildly annoyed that he could be so misunderstood. A better salary (which he thought the NYBG could probably ill afford), improvements in working conditions, even improved status, would have no influence on his decision at all. Surely people understood this? The only important thing to him was: would such a move be good for botany in general and world conservation in particular?

From this perspective he felt the directorship of Kew might very well be the right thing for him if it would give him some power and influence to use for the good of the planet which he felt was heading into environmental crisis.

On the other hand his established work at the New York Botanical Garden and in South America was obviously much valued. Trips to the rainforest from the UK would be much reduced. Perhaps this was still the best thing for him to concentrate on in the future. There was still so much to do there. There were several florulas, or local studies, he had planned for the Amazon, to follow on from Projeto Flora, there was work at Au Sable which he increasingly enjoyed and then his Institute of Economic Botany had several fascinating programmes just producing their first findings.

All rather lofty thoughts perhaps, but all genuinely and carefully considered and matched with Iain's coolly accurate, almost innocent, assessment of himself and his likely ability to contribute to each possible scenario. What, he thought, would someone of my talents and experience be best employed doing under the prevailing circumstances? His attitude was that of one disposing of an available, useful, botanical asset, not that of a man making a personal career move.

After some weeks of such consideration, he decided to accept the offer from Kew and agreed that his name might be put forward as the next director of Kew. The management of the NYBG accepted his final decision with good grace, despite their great disappointment in losing him. At this Iain was very relieved. He had never wanted to be regarded as indispensable, and the recent turbulence in New York had seemed to him as if one or two had already come rather too close to doing so.

Then followed a period of waiting while his name and credentials were forwarded to the queen for royal assent. This occasioned a strange mix of light-hearted curiosity and muted respect among his colleagues in New York. It was difficult for any in the world's greatest democracy to conceive of an appointment which had to wait upon the deliberations of the hereditary ruler for confirmation. But they loved it all the same. Many were the comments that Iain was daily waiting for a phone call from the queen herself, 'just to check out a few details'. Some believed that he had actually had one, though Mickey, with a gleam in her eye, would neither confirm or deny that this was the case. If she had had a tiara she would probably have fetched it out to wear in the office to add to the fun.

But Her Majesty evidently saw no problem in accepting the recommendation of her advisors without any such transatlantic telephone calls and appointed Iain as the director designate of her botanical gardens at Kew in January 1987.

In doing so she awarded him executive responsibility for the premier botanical garden in the world and a civil post rated at British ambassadorial level, by virtue of it being a royal appointment. In theory at least he could pick up the phone to Her Majesty at any time, where plant matters were concerned. In fact he was shortly to meet a good number of the royal family during their several visits to the gardens. Prince Charles, well known for his love of plants, is a great Kew fan and patron of the Foundation and Friends of Kew.

Iain also became instantly answerable to a broad and vociferous general public on all matters horticultural, and bearer of an historic mandate from botanical science which ran for over two hundred years.

He, as Kew's twelfth director, was joining a scientific dynasty of powerful men who had steered Kew to an almost unassailable reputation worldwide.

> 'I have more than three years bene a dayley wayter (on Lord Somerset) and wanted the chefe part of the day most apte to stydy, the mornynge... (but)... for these three yeares and an halfe I have had no more lyberty but bare three weekes to bestow upon ye sekyng of herbes, and markyng in what places they do grow.'

So, in 1551, wrote William Turner, a rather put-upon physician to the Duke of Somerset, Edward Seymour, then the most powerful man in England, being Lord Protector to the boy King Edward VI. William Turner was writing in the preface to his first *Herbal*, a book of plant descriptions. Though not the earliest of its kind, it was certainly one of the most thorough and well researched in England at the time. It earned him the title of Father of English Botany.

William studied in Pembroke Hall at Cambridge in the 1520s and became a preacher along with the two famous future Protestant martyrs Latimer and Ridley (Ridley taught him Greek). At the time he also took an interest in medicine but found to his annoyance that there was no consistent way ('in Greke, Latin nor Englishe') of identifying the herbs he should use for treatment. It was this that led in later life to his recording, naming and illustrating the many different types of herb he could find for his *Herbal*.

Although this was based on local field research he also grew many exotic plants and herbs in his own garden which lay near a little village just across the Thames from Syon House where his demanding master, the Lord Protector, lived. In the sixteenth century this village was called 'Keew'.

Though William Turner had no connection whatsoever with the establishment of a royal garden on the same site some two hundred years later, there was already a claim staked by botanical and even royal history on this low-lying, rather poor stretch of sand and gravel on the south bank of London's arterial river.

Elizabeth I was also supposed to have liked 'Kewe' (spellings vary widely in contemporary records). This was probably because her favourite, Lord Robert Dudley, lived there in Deyrie House, though no one is quite sure exactly where this house was. The traditional name of the bank-side by Kew's present Brentford Gate is Queen Elizabeth's Lawn which is believed to date from this time. Here Queen Elizabeth's Elm was planted, under which she is said to have sat and flirted with the Lord Dudley. Since she seemed to do this almost everywhere for most of her life (without ever agreeing to marriage) this seems highly probable. This romantic and venerable tree blew down in 1844 and its wood was made into a table for Osborne House, Queen Victoria's favourite summer residence on the Isle of Wight.

It was in search of an agreeable summer residence (and also to keep clear of his father, King George I, who couldn't stand the sight of him) that brought the first permanent royal resident to Kew: George, Prince of Wales and his princess, Caroline of Anspach. They purchased the Richmond Lodge Estate, which included part of the present Kew grounds, and hunted and entertained there extensively.

Princess Caroline, who became queen when her husband ascended the throne in 1727, was an energetic woman very free with her own and the public purse. These she used energetically, particularly the latter, to build up what then became her own estate. Her lifestyle was grand and eccentric, as were the constructions she ordered to be built in the grounds, which included a free-standing 'cave' with bizarre wax figures inside intended to celebrate the 'legendary Story of Merlin, our *British* Wizard'. She was perhaps trying hard to show the natives that the German dynasty then lined up for the British throne still had a high regard for indigenous mythology.

She did however allow the Richmond Lodge gardens to be opened to the public when the royal family was not in residence, a precedent for the future, and the various 'walks were full of company every evening to the great advantage of the town and neighbourhood' according to one chronicler of the time.

Queen Caroline died in November 1737 and the following year, true to what seems to have become a Hanoverian family tradition, her eldest son Frederick was denounced by his father, George II, as 'the greatest villain ever born'.

So Frederick promptly crossed from Germany, where he had always lived, to make his home in England, marrying in 1736 a fellow German, Princess Augusta of Saxe-Gotha, to settle with her and their baby (the future George III) in Richmond, just as his father had done before him. Far enough up-river from the City of London to be out of sight, but close enough not to be out of mind.

This time though he chose to live in the Kew part of the Richmond Lodge estate bringing up his family in the 'White House' – a residence now demolished but whose site is commemorated by a sundial in the present Kew Gardens. The Dutch House next to it is now known as Kew Palace.

Frederick and Augusta loved gardening and pitched energetically into the business of creating a distinctive home for themselves, taking the lead in the new 'natural' fashion for landscaping. This he left largely in the hands of professional architect, sometime designer of royal barges and landscape artist, William Kent.

Perhaps Kent's competence on land and water was so all-embracing that Prince Frederick felt a need to specialize elsewhere. In any case he began to take a more serious and direct interest in the plants which went into his gardens, making the acquaintance of the Earl of Bute, an avid collector of plants internationally – a very new 'hobby' – and also Dr Stephen Hale who had first demonstrated, in his book *Vegetable Staticks* (1727), the basic processes of plant physiology. He too lived by the river – at nearby Teddington – and was often visited at home by the enthusiastic prince.

Frederick was in the throes of preparing plans for a magnificent hothouse ('three hundred feet in length') at Kew, to accept and nurture some of his own and Lord Bute's recent overseas acquisitions, when he caught a severe chill and died, at Kew, in March 1751. England lost a future king, but the foundations had already been laid for the future Royal Botanic Gardens.

His widow Augusta, with help from Lord Bute who now became something of a surrogate father to her son, the future George III, proceeded to build on her late husband's vision and laid out further extensive gardens (in 1757) with the help of landscape artist and architect William Chambers.

In 1759 a new bridge over the Thames at Kew was given a grand opening by George II, who celebrated later with fireworks on Kew Green, while Princess Augusta employed a young Scotsman called William Aiton to have charge over her two 'specialist cultivations' within her more general gardens.

These were the Physic Garden, generally herbaceous but thought to contain a number of plants useful to medicine (William Aiton who had recently worked at the Chelsea Physic Garden was considered likely to know), and her Arboretum for trees and shrubs. These occupied between nine and eleven acres at Kew where the plants were set out in long rows at the end of which was a metal sign carrying the name of the planted genus, with each species subsequently numbered and referred to a manual called the *Herbaceous Book.*

Thus was recorded the start of a systematic approach to the collecting, growing and naming of plants at Kew. It is usually considered that the Royal Botanic Gardens were first formally established 1759 – though at the time they were no more than a private extravagance.

Shortly after this the most unusual non-botanical contribution to Kew was built, by William Chambers, working to Augusta's directions. This was the famous Chinese-style ten-storey Pagoda whose construction was started in 1761 as part of a group of 'round-the-world-architecture' buildings. But although more than twenty garden follies and other exotic buildings were constructed, only the Pagoda, three Greek-style follies (the Temple of Bellona (1760), the Temple of Aeolus (1760) and the Temple of Arethuse) and the Chambers' Ruined Arch bridge (1759), now survive. Chambers also built at this time, in the northern part of the gardens, the elegant Kew Orangery.

On the death of Princess Augusta in 1772, her son, now George III, inherited Kew. Occupying the White House officially, he also built his queen, Charlotte, a summer-house there in the style of a rural cottage. (George had pretensions as a farmer and was often dubbed Farmer George by his court.) It stands to this day, a delightful residence – as Queen Charlotte found it to be – still named Queen's Cottage and still set in a remote and quiet part of Kew Gardens.

George III also installed Sir Joseph Banks, recently returned from a round-the-world trip on *HMS Endeavour* with Captain James Cook, as his horticultural advisor, *de facto* first director of the gardens at Kew. Joseph was as much entrepreneur as botanist and had the required vision and energy to establish the gardens as a major centre for plant collections – some 7,000 new species being introduced to the UK during the reign of George III – and in 1789 published the first full catalogue of the collections, compiled mainly by William Aiton and Jonas Dryander (Banks' librarian), as *Hortus Kewensis* in 1789. Banks transmitted his enthusiasm round the empire, which equated to round the globe. Collections flooded in from Canada, Fiji, South Africa and Botany Bay in Australia among others. It was Kew's first golden age of discovery.

But although the king used Kew extensively (latterly moving from the White House to Kew Palace – which still stands in the grounds) and continued to admit an appreciative public regularly, when both he and Sir Joseph Banks died in 1820 the lack of subsequent royal patronage soon told against the gardens. By 1837 things had been allowed to slip so much that a committee was set up to review the state of the gardens with a view to breaking them up altogether. This committee, which included Joseph Paxton who had first flowered the giant water lily *Victoria amazonica* at Chatsworth, and John Lindley, eventually found in favour of keeping it the gardens together as a publicly funded institution. Political support for this expensive option came mainly from a number of influential but 'invisible' science lobbyists – which included titled amateur botanists.

But there was little enthusiasm at the Treasury for this decision, and funds were initially hard to come by. What was needed was not vague scientific influence but an individual visionary scientist to take charge of things. Fortunately one soon came along in the shape of Sir William Jackson Hooker, just awarded his knighthood for botanical research, following an intense career inspired by an early meeting in 1806 with Sir Joseph Banks. After this young William had 'determined to give up everything for botany'.

Sir William Hooker took over as official director of the Royal Botanic Gardens, Kew, now placed under the Commissioners of Woods and Forests on 26 March 1841. He intimated he would not need to spend very much government money at Kew and would keep matters there small, neat and undemanding. He had no intention of doing anything of the kind.

Over the next twenty years he skilfully ensured that Kew became nothing less than a controlled botanic explosion before the eyes of an increasingly plant-struck Victorian society, and made the gardens of such national interest that he and his Kew became, politically, virtually untouchable.

By 1845 Sir William had been given control of the adjacent pleasure grounds in addition to Princess Augusta's small garden plot and begun to spread the collection out. The Palm House, designed by Decimus Burton and completed by engineer Richard Turner, was built between 1844 and 1848 and became a monument to the impressive functional architecture of the time and the focus for his enlarged gardens. It was designed to do the job it does today: house trees and shrubs, but of a tropical kind never before seen on England's temperate shores.

Visitors were amazed to find themselves walking 'in the tropics' under soaring archways of palms and bamboos, and able to climb high among their leafy fronds on 'flying' (suspended) iron pathways. And all just a short ride from cold and foggy London on the exciting new railway train.

He enlarged the pond in front of the Palm House and called for the creation of a number of vistas and walks, to be undertaken by William Nesfield. Today these are still one of the most attractive features of Kew as a recreational garden. Here the Victorian public felt able to 'ape their betters' by strutting, picnicking or courting along shady avenues, past decorative hedges and neatly planted flower-beds like Lords and Ladies on their country estates.

Today the Palm House is still the focus of the main part of the gardens. From here towards the Orangery lies the Broad Walk with its decorative seasonal plants; close by is the famous Maidenhair tree (*Ginkgo biloba*), planted in 1762 for Princess Augusta. Broad Walk turns a sharp right angle to Decimus Burton's wrought iron Main Gate which leads out to Kew Green.

Looking to the south, the Pagoda with its bright red eaves can just be spotted through the trees, while to the west a long vista leads the eye out past the lake and the Bamboo Garden over the Thames to Syon House, the one-time residence of Lord Somerset and now the property of the Duke of Northumberland.

The Palm House was closed to the public in 1984 after 136 years of humidity within and British weather without had taken its toll. It needed a thorough refurbishment (every glazing bar and every pane of glass was replaced). The restoration included the installation of a Marine Plant Display in the newly excavated basement aquarium. It re-opened, still to most the visual centre-piece of Kew, in 1989.

Sir William Hooker also introduced the first supportive scientific institutions to Kew. In 1847 he opened the world's first museum of economic botany in a former fruit store; ten years later he had a new museum built for economic botany (once again by Decimus Burton) by the Palm House, and in 1862 converted the Orangery into a wood museum. The King of Hanover's residence became vacant on Kew Green on his death in 1851 and, it being hard by the Main Gate, Sir William quickly acquired it and converted it into a herbarium and library. He also acquired a home of his own, as Queen Victoria made another house available on Kew Green as the director's official residence – where the Prances live today.

In this there was something of a *quid pro quo*, as Queen Victoria, who took a great interest in Kew, had fallen in love with the Queen's Cottage and had the immediate grounds around it fenced off for her own use, insisting they be kept in a semi-wild and natural state. Though these grounds are now fully accessible to the public, her wishes have been strictly observed and the springtime carpet of bluebells in the wooded glades by Queen's Cottage is as impressive, and as natural, as the one she enjoyed so much during her reign.

Sir William added extensively to both library and herbarium collections in his lifetime and, on his death in 1865, Kew acquired his private collection of books and specimens. With George Bentham's earlier gift, these form the basis of the six million dried specimens available there to researchers today.

Sir William, like Sir Joseph Banks before him, authorized expeditions around the Empire and elsewhere, paying for plants collected and research done. One of the most notable was an expedition in 1860 to South America by Clements Markham, a respected field man of the time, and Richard Spruce who had spent much of his life in South America, particularly the Amazon. (Many have likened Iain Prance's work to his.) On their expeditions Spruce and Markham ascended the high Chimborazo mountain in the Andes and found and collected there seeds and seedlings of the Red *Cinchona* tree – the bark of which contains quinine. That same year these were sent out by Kew to India and in due course over a million trees were reproduced, saving many from malarial fever.

Perhaps the best indicator of the hold that Kew acquired on the public imagination with Sir William Hooker in charge was the overall number of people who visited. Such a botanical showman was he that from the bare 10,000 visitors a year that was the norm during his first years as director he eventually encouraged a staggering *half million* to visit the gardens in 1865, the year of his death. (It first topped a million in 1882.)

By this time he had also put in hand the construction of yet another vast glasshouse to be even larger than the Palm House, now known as the Temperate House. Though of more conventional rectilinear design this was not completed until 1899 to Decimus Burton's orginal design, and in fact proved rather unsatisfactory due to problems in ventilation and heating until renovated and somewhat re-engineered between 1978 and 1982.

The death of Sir William cast a giant shadow across the gardens and the British botanical community closely associated with it, but already there was someone ready and waiting to dispel the gloom and accept the mantle of directorship. His style would be different but the name was the same: Sir William's son Dr Joseph Hooker had already been assistant director at Kew for ten years. It seemed very natural to everyone that he should take the step up.

A field researcher and taxonomist, Joseph had already travelled with Captain Ross to the Antarctic (1839–43) and produced floras there and for New Zealand and Tasmania. Between 1847 and 1851 he collected again widely in Nepal and Sikkim. He was the classic roving botanical scientist and when he assumed the directorship immediately began to lay the foundations for Kew as a centre of science. Even as assistant director he had begun to publish, with George Bentham, *Genera Plantarum* – a now-standard work of background plant genera classification. Volume one was produced in 1863, and the three volumes completed in 1883 containing 7,585 genera and 95,620 species.

In 1876 the Jodrell Laboratory was opened (sponsored by Hooker's friend Phillips Jodrell) and in 1877 a further extension was added to the herbarium to house the rapidly growing collection much attended to by Sir Joseph. Field work continued – mainly in support of imperial commerce – with tobacco being planted round the Empire with varying degrees of success. In 1872 teak was introduced into Jamaica, tea, coffee and chocolate into Ceylon and more *Cinchona* into India while Kew-trained men were sent to Natal, Constantinople, Calcutta and Mauritius to botanize, collect and dispatch specimens back to Kew.

Back in Kew the business of formal horticultural education was at last addressed (instead of an informal working apprenticeship scheme) with Sir Joseph taking a leaf out of his father's book and wangling more funds from the Treasury to fund botanical lectures in 1873, the year he was elected as a Fellow of the Royal Society, possibly the most distinguished honour to be offered to a British scientist.

Early on the flood-tide of 14 June 1876 a ship slipped quietly up the River Mersea to Liverpool Docks. It was met by a special express train with steam up, ready to haul the secret cargo aboard off to London just as soon as it could be offloaded. Passenger Henry A. Wickham was acting as agent for the Royal Botanic Gardens, Kew, and the hiring order for the special freight train was in the name of Sir Joseph Hooker himself. The cargo was 70,000 seeds of *Hevea brasiliensis*, the Brazilian rubber plant, taken by Henry Wickham personally from the Amazon tributary Rio Tapajós.

From this shipment some 1,900 plants were propagated under glass at Kew and then dispatched by fast steamer to Ceylon. Enough survived to establish the entire Far Eastern rubber industry, possibly the greatest single feat of economic botanical management in history (if you discount the destruction of the rainforest – which is hardly 'management').

The gardens had been rather a male preserve ever since Princess Augusta's time, but in 1882 the Marianne North Gallery became the latest addition to the Kew buildings. It was built to house the almost magical paintings of Miss North who had travelled the world in order to paint plants in their natural habitat. 848 of them now crowd the walls of this gallery, which she herself designed.

Fourteen years later women were actually *working* in the gardens, though under orders to dress much like the men: in suits, thick socks and a peaked cap so as not to 'distract' the male gardeners from giving their prime attention to the plants.

Sadly, in 1912 the suffragette movement impacted on Kew when some women broke into the Orchid House and wreaked violent havoc there one night. For any women to have condoned such damage to so many rare and beautiful flowering plants is perhaps a vivid indicator of the strength of feeling then current in their movement for voting rights.

Though still very much alive and actively pursuing his research in the Kew Herbarium, and his friendships (he was a close friend to Charles Darwin, until his death in 1911), Sir Joseph Hooker handed over the directorship in 1885 to his son-in-law and assistant for ten years, William Thiselton-Dyer. Thiselton-Dyer took the gardens into the twentieth century. He too stayed for almost the regulation twenty years, enduring the limiting effects of economics introduced by the Boer War and the rise of the Labour Movement which soon highlighted the rights of his underpaid gardeners, then paid only twenty-one shillings a week. A labour dispute in his final year (sorted out by Labour MPs) helped hasten his departure and the post then passed to a former Indian civil servant Sir David Prain who was Kew director throughout the Great War.

He introduced a charge for entry – one penny – to assist the war effort, but this was removed by the Labour government of 1924. Charges then went on and off with successive changes of government, rising to three pence in 1952 and ten new pence in 1980. Today the entrance tariff is measured in pounds rather than pence, as the gardens no longer obtain full government support, but Iain feels the current charges are lower than other family attractions in and around London.

In 1922 Sir David Prain retired. Perhaps because of his background he had devoted much of his attention to promoting the production of the colonial floras. These had been first proposed by Sir William Hooker in 1857 and continued well into the twentieth century. Floras of New Zealand, the West Indies, Ceylon and South Africa were completed by 1865, Australia by 1878, Tropical Africa, Seychelles and Mauritius by 1877 and British India by 1883. It is a staggering amount of field and herbarium work. Many former colonies assumed responsibility for their own economic destinies in the 1920s so Kew involvement in running economic botany for the Empire reduced sharply after that, though links still remain strong with botanical gardens and universities in most ex-colonial countries through the mechanism of the Commonwealth.

Sir David was succeeded by his deputy Sir Arthur Hill who served the gardens long and well into the Second World War, but was tragically killed in a riding accident in the Old Deer Park adjacent to the gardens at the close of 1941. The senior economic botanist at Kew, Sir Geoffrey Evans, stood in temporarily until 1 September 1943 when Professor E.J. Salisbury of University College London was appointed director. He introduced a chalk garden – the first attempt at Kew to create an ecological community – and also, of necessity, supplied Kew with klaxons. This was an attempt to amplify air raid warnings to the brave glasshouse gardeners of possible V1 or V2 rocket-bomb attacks; for, working as they did beneath tons of fragile glass, these people stood to suffer severe injuries even from very distant explosions. In the event Kew was never directly hit by bombs though glass was smashed, as feared, by some attacks.

Professor Salisbury (who was knighted in 1945) retired in 1955 with the unusual Australia House, for uniquely Australian flora, also to his credit (this was built at Kew by that country in 1952) but his successor was to have more scope for change as wartime restrictions eased.

Sir George Taylor saw the need both for more space and more science and, advised by his eminent keeper of the herbarium and library Dr W.B. Turrill and a Kew visiting group of 'distinguished personages', he expanded both. He appointed a cytogeneticist and a plant physiologist, adding two new botanical disciplines to the gardens' scientific workforce which, along with a rise in student numbers, meant science floor space at Kew was now at a premium. The answer was the new Jodrell Laboratory and Lecture Theatre opened in May 1965, still the centre of the scientific community at Kew today.

He also added one more wing to the herbarium and library (the fourth since its purchase). He also made arrangements for Kew to expand its planting space by taking over Wakehurst Place near Ardingly in Sussex on 1 January 1965. This had been the home of Gerald Loder, Lord Wakehurst, former president of the Royal Horticultural Society, and already possessed many rare plants. It also had fine soil (unlike Kew), was located on a sandstone ridge and recorded higher annual rainfall. Still something of a Cinderella in some people's eyes – who want to think of Kew as just a historic garden near Richmond – Wakehurst Place has more than justified the hopes placed in it by Sir George, as available planting area, a centre for ecological research, and more recently, a base for a very new type of botanical collection: the Kew seed bank.

Sir George Taylor retired in May 1971, the last knight to hold the post, to be followed by Professor John Heslop-Harrison and Professor J.P.M. Brenan who served for five years each. The latter handed over to Professor Arthur Bell, Iain's immediate predecessor.

During their time life did not stand still at Kew, with new buildings being constructed and new scientists (and sciences) being approved and appointed. Constructions such as the Sir Joseph Banks Building, the Princess of Wales Conservatory and the Alpine House are all gleaming modern creations in light aluminium and glass, cunningly designed with protecting earthworks and sloping sun-catching roofs to maximize energy conservation – making them a statement in themselves – as well as providing exciting locations for their various exotic, international or unusual plant collections.

But by the mid-eighties times were changing more profoundly. In the new strident political climate Kew was being asked to re-justify its reliance on government resources. With no Empire to support and the modern public appetite for leisure satisfied by so many other attractions and the many stately homes opening their gardens to visitors, the old political arguments for Kew had worn a little thin. Though science and conservation had proved to be strong new ones, from now on Kew would be expected to find good reason for its existence on a continuing basis – and good money too; for not every project would automatically be supported by government paymasters. Forecast policy indicated Treasury funding was to be gradually reduced as, hopefully, Kew found alternative supporters. If not, then Kew itself would have to reduce.

Iain foresaw he was going to need all the experience he had gained in dealing with 'City Hall', national government and private sponsors in the USA to keep his new command under way.

20. GRASPING THE HELM

Having made the decision to go to Kew, the Prances began to prepare to leave New York. After twenty four years based in the city, going to live in London felt almost like facing a move to a foreign country. And leaving New York took some getting used to. Both Anne and Iain had many friends, Anne amongst her students of English and their families, Iain his scientific colleagues at the NYBG, not to mention their neighbours and church friends. There was strident dismay from every quarter at losing them. But they were sure it was the right move.

In the meantime they paid a flying visit to England to visit some of the senior staff at Kew. Iain already knew Kew well from his student days and had made many subsequent visits since - in fact virtually every time he had chanced to be in England. He soon learned to his delight that he had one or two very old friends on the staff - from his time at Oxford: Terry Pennington for one and Ray Harley, by now the world's leading expert in mints, for another. He was also particularly well acquainted with the magnificent Kew Herbarium, from his many visits there with Frank White. This is still the envy of the botanical world: Iain's NYBG colleague Dr Pat Holmgrun, giving her speech at a valedictory dinner in Iain and Anne's honour, said that she would be quite happy to see Iain leave New York if he would send her back the Kew Herbarium in return!

Anne did not know Kew, and was in two minds whether she should go along with Iain for his 'handshake visit' at all. She had not been particularly involved with the institutional side of Iain's work in New York, leading much of her own life - very busy with language teaching, her two teenage daughters and her home. Surely, she felt, Kew would prove to be little different? But in the end she decided to join him on the trip, if only out of courtesy to the Kew staff, Iain's future colleagues.

She was grateful that she did. Walking around the Gardens for the first time with Iain and Professor Bell, she became very conscious of Kew's sense of identity. Not so much as a beautiful and unique garden, or of its Royal place in history, nor even a sense of the work of the so many individuals whose joint efforts made up the garden's business. It was really Kew as a community which struck her. It seemed to her that whilst it knitted together well enough at a scientific or horticultural level (though she did not feel too qualified to comment on that), it ought perhaps to feel itself more of a body, a group of wildly different people but very conscious at the same time of the same aims and goals.

Everyone should feel they were 'Kew people' first, then scientists or gardeners, researchers, students, police officers (Kew Gardens has its own constabulary) or administrators, guides or caterers. She also felt that if this were to happen then the work of Kew would improve in all ways: scientific, horticultural, interpretive, and so bring everyone more job satisfaction.

Anne did not know it then, but this insight was to prove the foundation of her future role at Kew, and the basis for much of Iain's management style too.

So, returning to New York, they said their goodbyes and Iain took over formally at Kew in September 1988, and they began to settle in.

From the start, not unexpectedly, his biggest practical headache was money.

The 1983 National Heritage Act had made substantial changes in the way places such as Kew Gardens would operate. Until then Kew, as a government department, had been financed wholly by MAFF, the UK Ministry of Agriculture, Fisheries and Foods. But now increasingly such facilities were being expected to find other sponsors to help. Core work such as basic scientific research and the maintenance of the many heritage buildings in the garden would continue to be paid for by MAFF, but the use of the gardens as a public amenity, the mounting of expeditionary field work and specific overseas projects would now have to be paid for, certainly in part, from other sources. This might well include the then ODA - Overseas Development Agency, a government body (now the Department of International Development), but it was made clear to Iain that Kew projects would be judged alongside all the rest when it came to cutting up the ODA cake.

What brought all of this into particular focus for Iain at the time was the impact of the 1987 hurricane. On the night of 16 October 1987 winds of unprecedented force swept across the south of England, ripping up trees, blowing down small buildings and lifting the roofs off larger ones. There were a number of deaths. All seven of the proud oaks which had stood for hundreds of years in the market town of Sevenoaks in Kent, the county known as the 'garden of England', were bowled over like skittles. Row upon row of sturdy trees which had crowned the North and South Downs hills near London, or lined the 'Hogs Back' in rural Surrey or clothed the summit of homely Box Hill of literary fame just south of the capital now looked like the teeth of an unlucky boxer. A few were left standing, most had been smashed away.

In the morning, people in the home counties awoke (those who had managed to sleep at all) to find three quarters of their minor roads impassable. Without chain-saws many could get no more than a few yards from their own front doors. And such was the scale of the damage that roads remained blocked for days, even weeks to come.

But what seemed to disturb the nation more than anything else were the television news pictures, not of devastated Kent or Surrey, but of the Royal Botanic Gardens at Kew and the satellite gardens at Wakehurst Place in Sussex. Wherever the camera panned, tree after valuable tree lay torn and uprooted, sprawling carelessly across its upturned neighbour, gouging through cultivated grasses or flattening precious rockeries and beds of flowers. Trees which had weathered two hundred or more autumnal gales had been tossed about like broomsticks. Unique trees, trees from far away lands, commemorative trees, irreplaceable trees. The furious wind had shown no mercy, and spared no favourites.

Wakehurst Place suffered the most, its exposed southwest ridge and sloping hillside catching the full force of the westerly blast. Here some 95% of the trees over 10 meters high were lost or severely damaged, over 15,000 were lost in a single night on the 500 acre property.

And the British public, careless or at least stoical about its own gardens' losses was appalled at Kew's, and for many months afterwards Kew secretaries and telephonists were answering concerned letters and phone calls about the gardens or sending off notes of thanks to people who had given spontaneous donations to 'help re-plant all those trees.'

And Iain had joined Kew a matter of months only after the hurricane had struck. To everyone's credit he found that a great deal of the essential clearing up had already been done - certainly at Kew itself, Wakehurst Place naturally took a good deal longer - but that extra funding, a cash injection, was going to be needed to set the Gardens back on an even keel. And he was told that whatever disaster might befall Kew the new financial systems were firmly in place. Yes, there would be some extra government money made available but Kew would now be expected to play its part too.

Here Iain's long term experience in the USA stood him in excellent stead. Even as a scientist he was quite used to talking about money, having spent many years thinking of projects one day and wondering how to fund them the next. The Institute of Economic Botany, the 'baby' which had been given him in New York had needed close on two million dollars to set up. And he and Jim Hester had simply had to go out and find it themselves.

So he set about his task American style - a combined one of educating the Kew staff internally in the business of promoting 'the plant business' and going outside to get possible sponsors to take an interest in the gardens. He orchestrated the composition of a Kew Mission Statement and Corporate Plan, called meetings, held working breakfasts lunches and dinners, called on ministers and potential sponsors and generally put Kew on a lot of influential agendas.

The working breakfasts did not last long. The traditional British enjoyment of silence at such a meal, excepting perhaps for a polite request to pass the marmalade or the daily newspaper, seemed to outweigh even Iain's notorious ability to promote early morning enthusiasm. But dinner was to prove a different matter. The working meal at the evening hour soon led to the establishment of something that has now become an institution - the frequent 'Director's House Dinners'; organised, hosted and often cooked by Anne Prance herself to which international dignitaries, industrialists, diplomats and the occasional prince or princess are invited.

On their preliminary visit to Kew, Anne and Iain had decided that the large Georgian Director's House on Kew Green had more living space than they needed, so, taking themselves upstairs, they turned the ground floor into an exclusive hospitality suite on behalf of the gardens. Here, several times during the week, anyone who might be interested in the work at Kew are invited to meet some of those who work there, and to discuss with them the modern role and purpose of the gardens, with Iain and Anne acting as personal hosts.

Out of these soirees or dinners, at which Iain or one of his colleagues usually makes a brief explanatory speech or shows a Kew presentation, has grown a wide interest amongst many hundreds of influential and concerned people, who, first invited as guests, have returned as supporters and sponsors. Perhaps to underwrite a new Kew expedition, finance a research project or provide capital funding for an on-site facility. Also, Anne has found, it has helped the Kew people who are invited (from all parts of the organisation) feel that, whatever their work, they can usefully represent the gardens to the outside world. Many of whom would never have thought to have done so. As she felt when she first visited the Gardens: if the people of Kew become aware of the wider significance of their individual contributions, and also how much outsiders are impressed by and interested in it, they will become better at doing it and more fulfilled by it.

Sad though she was to give up her teaching, which she had hoped to continue in England, this vision of hers has shaped her own role and involvement in Kew and also enabled her to help many with their private problems and concerns. She has in this way helped keep the 'family' of the gardens together whilst her husband attended to the institutional fabric.

Another key figure at these events, indeed a man who until recently ran a number of similar functions of his own, has been Giles Coode-Adams, former Chief Executive of the Kew Foundation, the fund raising arm of Kew, another of Iain's innovations, who retired from this in 1997. Giles Coode-Adams, says Iain, is someone who has been pivotal in the work of the Kew Foundation, especially in his quietly detailed efforts, often unsung, which have proved authoritative and therefore crucial to progressing so many vital projects. Iain knows him to be a man with a whole heart for Kew and all it represents, who nevertheless adopted a keen business-like approach to the whole matter of fundraising for projects. A personal mix which proved formidable. Illustrated by his crowning fundraising glory: achieving a staggering £60 million to fund The Millennium Seed Bank project at Wakehurst Place.

So, through making many contacts and visits of all types, arranging seminars, giving speeches and leading appeals Iain and his colleagues at Kew managed to raise 50% of the money needed to commence restoration of both gardens after the 1987 hurricane; the other 50% was supplied by the government. Though as someone at Kew said afterwards, rather cynically but perfectly accurately: 'It's an ill wind that blows nobody any good - the Wood Analysts on staff thought Christmas had come early that year - and for most of the next!' It is certainly true that every opportunity was taken to conduct analytical research on the trees which had been uprooted at Kew in October 1987.

And in the Kew shop by Victoria Gate there is now a dramatic wood mural depicting the night of the storm in Greek legendary style. The wind is represented by a human figure blowing blasts of air with his mouth at the gardens shown as trees, plants and animals fleeing before him. In defence are two ferocious lions. The artist, Robert Games, just sixteen when he created the mural, had visited the gardens shortly after the storm and seen a fallen Turkey oak lying between two stone lions by the lakeside. They had remained untouched despite the ferocity of the storm and the carnage all around. The mural is made of over 1000 carefully cut and polished individual pieces of wood, of different textures and colours, taken from thirty three different tree species struck down in the gardens during the storm. And now, starting at the mural, visitors to Kew can now follow an 'after the storm trail' round the gardens to see remaining, living, specimens of some of the trees used. But they would be hard pressed to see any signs of storm damage. As Iain says: "If we didn't have this storm trail and mural you'd never know there had ever been a hurricane. A testimony to the extraordinary work of the staff and the wonderful regenerative power of nature."

The early chasing after funds at Kew was certainly not tied to restoring wind damage. It led to the introduction of a realistic, rather than nominal, gate tariff to cover some of the horticultural running costs in maintaining the living collection of over 30,000 species, the establishment of the Kew Foundation under Giles Coode-Adams, and the formation of Kew Enterprises to run the Kew Shop, building rentals, concerts and other commercial activities of the gardens.

The most flamboyant of these are the summer jazz concerts where enthusiasts, both single and corporate, can picnic in the gardens on a (hopefully) warm summer evening with the Temperate House as a back drop and world class jazz musicians performing on stage to the clink of glasses and the popping of champagne corks. The showman in Sir William Hooker would have loved it, especially the now mandatory firework display at the end. And the great thing about glasshouses at night is that they reflect the light - so Kew can offer quite a unique pyrotechnic experience and (though one or two might argue the point) the noise doesn't seem to frighten the plants!

In his first years as Director Iain also sought to establish a moral as well as financial support for Kew, by creating the Friends of Kew organisation which he started in 1990. Today this boasts over 40,000 members who, for a small annual contribution, attend meetings and other events at Kew and are kept in touch with Kew projects and news by their own quality magazine called simply 'Kew'. Iain comments that the remarkable success of the Friends of Kew and the quality of their magazine is largely due to the publishing background and organisational skills of their most effective manager, Michael Godfrey, whom Iain appointed to develop his Friends of Kew idea in 1989. Once again Iain felt that the sense of identity of Kew and the importance of its work could only be strengthened by a body of informed enthusiasts. And so it has proved.

Kew also publishes 'Kew Scientist', a non-technical monthly news-view of the work of all departments (not just Science) and 'Kew Guardian', the staff in-house magazine for all 560 employees as well as for members of the 100-year-old Kew Guild the association for existing and former staff and students.

Anne recalls an interesting conversation with her local Vicar during her first year at Kew. "I had said to him that I was now always so busy at Kew that I really felt badly that I could not help out and support the church as I had always tried to do wherever we had been before. He looked at me and said simply 'my dear, your ministry must be at Kew - and we shall be supporting you!' It was what I had felt. But that was a confirmation for me. And for us both in a sense. Iain and I were at Kew - and there together to be about God's business - in our different ways.

21. SCIENCE AND APPLIANCE

Some time later Iain codified the corporate style which Ann and he had sensed into the management structure. He also stated in his 1994 projected 'Vision for Kew,' (a five year plan for the Gardens) that he specifically wanted to see more interdisciplinary work done at Kew, with projects which stretched across traditional departmental boundaries. Multi-disciplined teams working together towards a valued goal rather than each department working independently trying to prove itself superior to the others.

Iain used the familiar Brazil nut tree to illustrate this in his presentation to the staff: it needing a combination of bees to pollinate it, orchids to provide 'homing' scent for the bees, Agoutis to disperse and trigger the seed germination, and mycorhizal fungi to assist the uptake of nutrients from the forest. This was the picture of Kew that he wanted then and one he now feels is more fully established, a continuing vision he wants to hand on to his successor. As one senior scientist said recently "The aim of Kew now is much more the running of successful programmes not just successful departments." And for the head of a powerful department to say this, shows that Iain has in good measure achieved his goal.

This programme-centred approach, coupled with Iain's global view and unusually wide personal contacts in the international botanical world resulted in many outsiders being drawn into Kew programmes. And Iain has taken great pains to ensure that such involvement does not simply bolster Kew and its reputation, but that credit is given internationally where credit is due. He has been particularly rigorous in his demands for ethical standards in all botanical dealings with other nations and has set up a legal department at Kew to advise on this, so Kew in turn has received outstanding co-operation from other countries.

A venture which classically illustrates this beneficial internationalism is the development of a wholly new Plant Classification system for flowering plants. This thunderclap in the taxonomic world was contained in a very modest looking document released in December 1998 and, strangely for such a profoundly ground breaking paper, it came out un-authored. being issued under the generic authorship of 'The Angiosperm Phylogeny Group' . It was presented by three named compilers, one of whom was from Kew, with around forty other assorted contributors spread right around the world. Most of these had never even been to Kew let alone worked there. It was a triumph of international co-operation.

Much of the re-classification evidence came from Kew's DNA systematics programme. Surely some noted scientist there should take the credit? Ot at least for Kew? Dr Mark Chase, Director of Molecular Systematics at Kew's Jodrell Laboratory, says no. That is the whole point. Plant classification must now get away from the business of one scientist's opinion set against another, or one botanic garden's reputation over another. He feels that his and others DNA systematics programmes now gives all plant taxonomists a base line for identification which was not previously accessible. Genes are after all the blueprint for life itself, he points out. Classify according to DNA and you are likely to be on the right track.

So, an international initiative with significant Kew input has now led to a worldwide revision of plant classification. And a great empowerment of plant science. Kew has not so much led the way as encourage others to move forward with it. The right way of conducting modern science according to Iain.

Interestingly, the whole of the Jodrell's development of these advanced, high speed methods of taxonomy using DNA sampling is something which Iain fought for soon after he came to Kew. That he did so was a surprise to many. For molecular systematics is something few thought Iain would look kindly at when he took over as Director. He was a taxonomist of the old school, a champion of visible detail, personal conqueror of the long slow climb to the top of his profession. Master of practical minutiae; a man of microscopy, tweezers and dry rustling leaves, who looked back with reverence to the impressive figures of Sir Joseph Banks and the assembled company of Victorian plant hunters. Surely Iain above all was the traditional taxonomist? He had, on his own admission, walked firmly away from the temptations of novel DNA research at the start of his career. Surely this leopard would never change his spots?

But Iain remembers the increasing certainty he felt in his career that DNA molecular sampling was to be a way forward in the future. Not a science for himself to practise perhaps, but certainly one to be embraced by botany as a whole, and particularly by Kew.

Matters came to a head following the first five year government scientific review that Kew had had to submit to shortly after Iain took over. A panel of external scientists were required to report on the work of the gardens and to outline a way forward: To say what, in their authoritative opinion, it was appropriate for Kew to be doing and what it was not. When this report was presented to the Kew Board of Trustees it effectively recommended that frontline plant science (such as DNA research) should be separated from the more prosaic business of plant propagation and collection, with only the latter being the proper business of a botanic garden.

The Board were about to accept this when Iain suddenly waded in. He made a powerful presentation proposing that a multi-disciplinary approach based wholly at Kew would be far the best way of achieving research results. A cutting edge molecular sytematics laboratory on site at Kew, would, he argued, link naturally and easily to both the living collections and the herbarium where it could use them as a direct source of supportive data on the doorstep. Not so, Iain argued, in a university (where the review committee had proposed such molecular research take place). For even assuming the very closest personal and organisational links - no foregone conclusion - any university laboratory would still have a great deal of further liaison work to do to confirm their results at one botanical garden or another. And there was also the matter of the invaluable support and 'unconscious' assistance gained just by working at Kew and meeting other plant experts daily, informally, and discussing the subject closest to their hearts. As Iain had found in New York, mid morning coffee time was often when one person's comment to another sparked off a whole new train of ideas.

Iain's appeal carried the day and Kew committed itself to molecular systematics and was soon building an extra wing on the Jodrell Laboratory to accommodate some 20 scientists and students with a DNA Sequencer at its heart. This resembles nothing so much as a coffee vending machine and is about as inspiring to look at. But it certainly cost a good deal more and what it does is little short of amazing: It can read the line up of DNA in any gene sample presented to it, recording and mapping the unique finger print of any species. Matching similar finger prints gives an indication of both family and near relatives.

When they got to work the molecular systematics team, under Dr Mark Chase, began to deliver some unexpected results. Not only could they match up some unsuspected 'sisters' and 'cousins' in plant families but they could speedily re-classify many which had been the cause of confusion and often controversy for many years. Soon, far sooner than they had dared hope, the need to revise plant classification became more and more obvious. Others in the same line of research were contacted round the world. They too were beginning to agree a change was needed.

As Dr Chase put it: "Before this if you were to go to six different text books on plant taxonomy and ask what is the nearest relative of the rose family you'd get six different answers because each author would emphasise a different component of the information. One may say the chemistry is more important, another the way they look and so on. They can't all be right! And of course some connections are more useful in one context than another. But to really know shoots science forward a quantum leap. People can then start making all sorts of connections they never suspected. Even an encyclopaedic knowledge of many species may not show up quite obvious links simply because no relationship is suspected. But knowing a relationship exists, (indicated by the DNA) gives any research a great deal more power."

Iain's interest in the systematics work at the Jodrell grew particularly keen when Dr Chase swung his DNA big guns in the direction of *Rhabdodendron*, the Amazonian bush which Iain had described as a new plant family (RHABDODENDRACEAE) when he split it off from the CHRYSOBALANACEAE early in his career as a student. Iain had contended that its nearest relative had to be the PHYTOLACCACEAE, due to the structure of its wood. But his widely respected, and much more experienced, taxonomic colleague in the NYBG, Arthur Cronquist, had disagreed. He argued for RUTACEAE - as the flower structure was evidently similar. It was the classic unresolved difference between scientists.

And so it stayed until the Jodrell put some *Rhabdodendron* material through its DNA sequencer. Out came the identifying string, and Iain had the last laugh. It was related to, though not within the PHYTOLACCACEAE.

Iain had been proved right, at least in that case. But it is a delicate business. Many scientists are understandably hesitant about letting the molecular sampler loose on their own taxonomic life's work. What if the DNA undermines many long cherished personal classifications? But Iain, says Dr Chase, has no such qualms. The truth is what he wants and his instructions are not to spare Iain's blushes.

As *Rhabdodendron* shows an interesting observation coming out of the DNA systematics programme is that obvious characteristics such as flower size and shape, for example, can often be very poor indicators as to family relationships. Flowers seem more often specialised to attract specific pollinators - insects, birds, etc, rather than giving clear indications as to family affinity. And the DNA strings have shown up some quite unexpected family relationships, that the cabbage is related to the papaya for example. Absurd, surely? But when a relationship is indicated other things often start falling into place. It has now been remarked that both the papaya and cabbage have a high mustard content, no doubt because of their familial connection. Other correlations are even more intriguing and have encouraged doubters to query the reliability of the DNA sequencer. The plane tree has been placed close relative to the lotus for example, which had always been thought to be related to the water lily. Apparently not. And there is now considerable agreement on this. But do the lotus and plane tree have anything really in common? There is certainly nothing obvious. But Dr Chase is philosophical. "I have sufficient confidence now" he says "that I just put the information out for discussion if the DNA gives that kind of indication. So far it has proved a reliable guide. As a scientist I have to be ready to be wrong, but my bet is that when someone takes a real good look at them both they'll find something which ties them up! You must remember that because something is the closest relative, it does not have to be a close relative"!

But if much of the science at Kew is highly specialist, relating to the investigation of plant chemicals, the micropropagation and study of obscure species or even DNA sampling, there is often a common application of such science which immediately touches society. One of the most dramatic of these is the accurate identification of poisonous plants in an emergency.

Every day of the year a Kew scientist with a taxonomic background is on call, by mobile phone or bleeper. He or she is the 'Duty Poisons Officer'. If by mischance someone is admitted to a casualty department at a hospital anywhere in the UK having swallowed a plant poison - deadly nightshade (*Atropa belladonna*) - for example, then the Kew Duty Poisons Officer can be called immediately for his or her advice.

This can be an emotionally taxing business for the scientist who must perhaps deal with a distraught parent of a young child who has returned from playing in a field or hedgerow and subsequently shown signs of having been poisoned. Rushed to hospital the child usually has no clear recollection of what they have eaten, or is unconscious, and assumptions may be made about pills or paint fumes or gas, until it dawns on the medical staff or parent that a plant poison might be the culprit. Police and parents will then have to try and retrace the steps of the child, who is possibly by now very ill, and take the advice of the taxonomist on the range of possibilities or, on their instructions, perhaps actually collect suspect plant material.

Fortunately a wise and experienced taxonomist working with quick thinking medical staff can usually narrow down the plant possibilities very quickly and young lives are saved. Kew is contacted up to two hundred times a year for this kind of taxonomic advice. Such was the significance of the problem that a scheme was set up by Kew in conjunction with Guy's Hospital and later expanded to include twelve others to introduce a little amateur taxonomy into hospital casualty units. This comprises a basic taxonomic data base of poisonous plants held on a computer with good visual graphics. A concerned doctor then works through a critical path of questions and answers carefully thought out by Kew taxonomists to narrow down the plant type by applying the available information. Pictures of the plant appear on the screen and full background details are supplied on associated pages concerning the relevant poison and antidote.

This project has proved a great success and has received an information technology award from the British Computing Society. Not only that, and far more important, medical staff report a 96% 'certain or nearly certain' confidence level in the identification of poisonous plants in this way. A good example of the practical appliance of Kew science.

Much has been made recently of the business of carbon dating ancient artefacts made thousands years ago, but Kew also has a practical handle on the distant past - the identification of 'period' pollen grains. Pollen grains can illuminate not only history but geography too. Nearly every ancient manufactured article has, trapped somewhere inside it, pollen from the age and location of its manufacture. From glue in old furniture to paint or varnish on the surface, wax in statues and filaments of cloth or sacking used in shrouds or to wrap up weapons or tools. To one who knows pollen (and it is a very complex subject) every particle tells a story.

Recently scientists at Kew were approached by a museum keen to acquire a Leonardo Da Vinci sculpture. But there was some doubt as to the authenticity of the work. Could Kew help? The palynological laboratory in the herbarium thought it could and scientists there painstakingly scraped away the base of the statue to find a wax plug inside the hollow form. The wax yielded grains of pollen which, when studied, revealed themselves to be from plants which would have grown in the fifteenth century in the North Mediterranean region. Whilst this was not conclusive (and Kew were careful to point this out) their findings were added to the weight of documentary and circumstantial evidence available to the museum and eventually persuaded them to purchase the statue as a genuine work.

Kew helps the fight against possible crime in other, less exclusive areas than the world of art. Scientists are on call by both the British Police and HM Customs and Excise to identify 'illegal plant based substances' - drugs - which they have seized, and to estimate their purity and country of origin. The type of drug 'mix' and the origin of the plant material can often give a good lead to the authorities as to the secondary source the drug - perhaps even the individual trafficker, as each one tends to have their own style of operation and drug 'signature'.

Kew scientists and horticulturists also supply consultant services to government departments for a wide range of projects including the evaluation of environmental impact studies on areas planned for new roads or buildings, forecasting the effect of pollution damage on flora when an environmental disaster strikes (such as an oil tanker spill) and outlining aims and goals for conservation policies in the UK, such as the UK Biodiversity Action Plan.

One of the most significant changes in the gardens themselves which Iain has achieved is the broad move towards conservation methods of the 'you-do-as-I-do' variety to show the domestic horticulturist and to some extent the private home gardener how this can be done.

Consequently the harmful use of peat as bedding for flowering plants and in the greenhouses has been stopped and a natural and widely available substitutes such as coconut fibre coir are now used. Use of chemicals has been minimised and in many cases stopped altogether. In the Palm House and other enclosures a carefully balanced ecological cocktail of ten insect predators, two fungi and a bacterium added to the internal ecosystem has meant chemical sprays are now not needed to regulate the (undesirable) insect population and prevent disease. This system called integrated pest management, is working well in the large public glasshouses of Kew and in many of the behind the scenes nursery glasshouses as well (recently fully rebuilt), and work continues to investigate ways in which it can be used in the outdoor areas too.

All of which organic effort makes one wonder about Iain's view of GM foods. What some say is a top example of plant science mis-applied. And something which many would consider dangerous ground for him to tread on in his position. Especially as Kew is in the business of attracting funding for so many different farming related projects. But, as ever, he is happy to comment without fear or favour, if asked . And asked he certainly is.

"Wherever I go to lecture now" he says, "and it doesn't matter what I'm there to talk about, someone will always bring up genetically modified foods. It has become of great concern to many people."

So perhaps some would be surprised that, given his convictions, both scientific and spiritual he is not dead set against genetic engineering on principle. "As cocaine can be a useful medicine or an addictive drug and atoms can be split for good or evil purposes in the same way genetic engineering is, in my opinion, a science which can be used for furthering the well-being of mankind - or it can be abused." He cites the case of insulin used to treat diabetes and other ways in which DNA technology is being used to produce medicine or treat genetic disorders.

But he has greater reservations when genetic work is used to produce ordinary food merely as a labour saving device or to increase profits. "The risks being taken may be too great" he continues. "Isolating and transferring one gene to an alien organism, to enable the crop to resist weedkiller, is a good example. Who is to know what effect this will have on the biodiversity of the farmland? Many species of insects birds and other animals depend on species of weeds. Or supposing the alien genes of a modified crop escape into the wild plant population, or into the crops of an organic farmer? This could happen all too easily, because plants are fertilised by pollen which is carried through the air, often for long distances."

But if this is a significant concern to Iain something he sees as much more directly repugnant is so called 'terminator gene' technology where crops are altered so that they cannot produce the next generation of seeds. The commercial advantage to seed companies is obvious, but in the third world such crops would be devastating to small farmers who depend upon harvesting some seeds to start the next year's crop. Iain says "it is not surprising that India has already banned this technology". Nor is it true that such technology benefits third world farmers by offering better yields, as the GM promoters argue. For their main aim seems to be to produce new products for western consumers, rather than to feed the poor. Iain feels that it would be better to concentrate on restricted applications of this technology which confer strong ecological benefits. But he also feels, nevertheless, that it is most unlikely that the current modifications are harmful to the consumer, whilst welcoming full labelling to give the consumer choice.

He comments "like all powerful technology let us try and urge governments and companies to use GM technology for good...not greed."

22. CONSERVATION AND COUNTING...

The theme of conservation, a fundamental to Iain's life, came to occupy Kew more and more as Iain steered it in that direction. And Iain's domestic profile as a serious scientist and conservationist began to be recognised. He was increasingly asked to talk and broadcast on issues, and occasionally about himself, including a particularly fascinating 'Desert Island Discs' on BBC Radio 4. His work was recognised in and out of botanical circles, overseas most notably by the award of the International Cosmos Prize from Japan in 1994 and in the UK most honourably by the award of his Knighthood - 'for scientific services to conservation' in 1995. Though the greatest honour, from Iain's point of view, was his election as a Fellow of the Royal Society in 1993, the highest scientific accolade which can be awarded a British Scientist - and especially felt by one who had spent most of his scientific life out of the country!

But prestige has never taken Iain's passion away from the field. Particularly if it a rainforest one. And, though he has often found himself tied to his desk, his love for the Amazon has not diminished and at least once a year he has managed to get back to Brazil, not only to see Rachel in Recife, but also to visit his pet project there: the Ducke Forest near Manaus. This has been a study in detail of the flora of one small, 100km square, reserved area of rainforest and is one of the florulas on which he had to hoped to work had he continued at the New York Botanical Garden. It has just been completed and published, though as Iain says with some understatement, 'the work is done, all it needs now is translating from the Portuguese'. To a man who is happy to slip back and forth from one language to another this might not seem too much of a problem but to one or two others from his New York days it still means 'work still in progress'.

The Projeto Flora Amazonica, which ended in 1987, had thrown up a number of future possibilities for florula studies in the Amazon which Iain fully expected he would have to abandon. But his personal enthusiasm for the Ducke Forest project seemed to create its own interest within the then ODA (Overseas Development Agency, now the Department of International Development: DFiD)) who were increasingly becoming interested in supporting plant conservation projects as wise long term investments. It wasn't long before the Ducke Forest Reserve survey was a Kew promoted, ODA-funded project with two Brazilians and a Kew scientist working on it full time in Manaus. And for twelve years now 'Reserva Ducke' has been Iain's first port of call after INPA whenever he lands back in the Amazon.

He is also reminded of the Amazon when he hears occasionally of the success of the travelling Margaret Mee Exhibition. Margaret Mee, an expatriate English woman who lived for many years in Brazil, was a celebrated painter of Amazonian flora. After surviving many adventures in the Amazon, she died tragically as a result of a car accident in Leicester in 1988. Iain campaigned for sixty of her Amazon flora paintings to be purchased for Kew. These are now on display in a roving exhibition hosted by various botanical gardens and horticultural societies to raise funds for the Margaret Mee Amazon Trust. The trust supports scholarships to help young Brazilian botanists, artists and conservationists to visit Kew and other UK institutions.

Iain's concern for the rainforest has, of necessity, widened beyond the Amazon and variously funded Kew projects address a number of rainforest conservation programmes worldwide. In Madagascar where rainforest clearance has reached epidemic proportions (only 25% of the original rainforest now remains) an extensive Kew survey to record all the species of palms that grow there, before they are lost, has been underway since 1992. For such a physically large plant one might imagine most palm tree species had already been well described worldwide - but Kew palm experts Henk Beentje and John Dransfield found no less than seventy new species of palm in Madagascar. The fast food chain MacDonald's sponsored this project - a company which Iain has confirmed has never been involved in the destruction of the rainforest despite much adverse publicity over the years.

In Irian Jaya the Western half of the island of New Guinea), where most of the land area - the size of France - is covered with rainforest,

Kew expeditions are again surveying and collecting in classic fashion to build up the basis for a full Flora. Their latest SE Asian expedition made 1,300 species collections there, some, wholly new to science, are so surprising that they have not yet even been placed in a family. And in Africa (a continent of long term interest to Kew - rather as South America has been to the NYBG) DFiD and Government of Cameroon have funded the Mount Cameroon Project - in an area widely considered to be the richest tropical rainforest in Africa.

Here over 9,800 specimens have been gathered so far with over 50 species found to be new to science - with much of recent collections still to be identified. There is also an increasingly significant study project here on Forest Regeneration based at Limbe in the Cameroon.

For Kew, and for Iain, the rainforest has still been only part of the picture. Various projects have also been instituted in Kenya forming part of a 'Flora of Tropical East Africa' compilation being run by the National Museums of Kenya in conjunction with Kew and supported by the global environment fund of the World Bank. And, at the other end of the rain scale, a Fuel Conservation Workshop in Harare recently studied the effects of over-cutting of wood for domestic fires in the semi-arid areas of middle Africa. New forest management techniques in these areas are being urgently researched where the thick brown dust and parched earth could not form a greater contrast to the green lush forest being investigated by Kew scientists on the Etinde Reserve in Camaroon.

In contrast to the New York Botanical Garden, which focuses its economic botany research on rainforest areas, Kew concentrates on arid regions of the tropics. For example Kew's flagship programme in South America - PLANTAS DO NORDESTE - is helping the arid semi-desert region of north east Brazil. (The area where Iain's eldest daughter now works.) Back in the UK, field project sponsors are also forthcoming. One very appealing, and successful, operation has been the Sainsbury Orchid Conservation Project Species Recovery Programme to save the British Fen Orchid (*Laparis loeselii*). Seeds of this delicate plant have been germinated at Kew using a symbiotic fungus in the micropropagation laboratory.

These have now been returned to the Norfolk Broads area, its native habitat, re-investing the sites where it has died out or is on the verge of doing so.

The British National Trust is also working with Kew on another similar project. The orchid *Orchis laxiflora* is now being re-introduced into

Jersey the isle famous for its summer 'Battle of Flowers' a carnival of flower covered floats and floral displays. It has also been successfully introduced to a meadow at Wakehurst Place. Both of these are part of the UK government Biodiversity Action Plan, in which Kew has played a central role. Sometimes, though, sponsors have simply given to brighten the gardens themselves. A particular favourite with both Iain and Anne (remembering her many English students from Japan in New York) was the rehabilitation of Chokushi-Mon, the Gateway of the Imperial Messenger. This, a four-fifths wooden replica of the Japanese cermonial gateway in Kyoto, the old imperial capital of Japan, was presented to Kew after featuring in the Japan-British Exhibition of 1910. It is regarded as the finest example of traditional Japanese construction in Europe. It had suffered much from weather and pollution and by the early nineties was in need of complete restoration. Iain authorised this having found both Japanese and British sponsors. It was completed in 1995 and surrounded with a Japanese landscape composed of special areas, most notably a Garden of Peace, crafted like a tea garden with lanterns and a dripping water basin. There are also Gardens of Harmony and Activity reflecting these moods containing uniquely Japanese flora and garden culture. Iain and Anne were especially pleased to have the renewed gateway opened by Princess Sayako herself.

As many of these examples show - and these are just a very few of the many Iain has been responsible for during his time at Kew - species conservation by re-propagation and preservation is a familiar and urgent business for botanical scientists at Kew. In 'micro-prop,' a research arm of the Living Collections Department, up to 1,000 endangered or expiring species may be under test-tube propagation at any one time.

But at Wakehurst Place in Sussex there grows, its buildings completed by 2000, one of the most ambitious projects of conservation which Kew has ever contemplated. Launched in May 1996 by Sir David Attenborough and described by him as "perhaps the most significant conservation initiative ever taken" it is called The Millennium Seed Bank.

This was a vision which grew out of the awareness that in the dryland areas of the world species were becoming extinct at an alarming rate. Kew's special interest in these areas was providing ample warning of this. It is a shocking statistic that some 25% of the world's flowering plants face extinction in the next fifty years because of changes to their habitat caused by forest clearance, desertification, unsustainable farming practices, climate change, building and industrial development and pollution. This was allied to the knowledge that Kew had developed unique expertise over a number of years to preserve seeds, at least from this kind of dry land area, in a suspended state, but capable of ready germination. It was called seed banking. At Wakehurst Place in Sussex there has been a seed bank for some time with seeds being stockpiled in a small way over perhaps twenty years. By 1990 the bank already contained some 2 million seeds, perhaps 2 percent of the world's flowering plant species.

Scientists at the Jodrell Laboratory at Kew and at Wakehurst had developed techniques for drying seeds so that they contain only about 5% moisture. Then they froze them. With such a low moisture content the seeds do not ice up and die, but can be preserved, in suspended animation, for perhaps two hundred years or more. As part of the storage process sample seeds are taken out of their long hibernation every five years and germinated to confirm they are still viable. Dry land seeds are particularly suitable for this process, for the more moisture that has to be removed to get down to the critical 5% the less is the likelihood that seeds can stay alive during the storage process. So wetland and rainforest plants are the most difficult to preserve in this way. Recognising that the science of storage for the dryland seeds was in place and that the conservational need was there - what could Kew, do to help meet the problem?

The answer was a vision to build a brand new seed bank as a base both for storage but also research in collaboration with countries round the world by the year 2000. A centre which would aim to go on to match the advancing century percent by percent in seed collection. By 2010 - ten percent of the world's species seed banked, and by 2020, twenty percent. The idea being to provide nothing less than an extinction safety net for the world's plant population - and from an area where human need, and sometimes greed, was causing the most damage. For in the dry lands plants are usually such a crucial part of a subsistence economy for the people living there that conservation is ignored out of desperation. In countries like Mozambique or Zambia, Sudan or the Russian Steppes. And when you are close to starvation, or need immediate clothing or warmth, all of which plants provide, conservation seems too long term a goal, even if you have been made aware of it. The Millennium Seed Bank was a daring, not to mention expensive vision - initial forecasts were for tens of millions of pounds. But Iain felt it was possible. It not only fitted in with his strong conservation statements on the role of Kew, but had a truly millennial dimension. "Preserving what God has given us" he comments "is my view of conservation, as most people know. So what could be more in keeping with the Christian celebration which the millennium should be than a conservation project of such potential?"

And possible it has proved. Five years later the Millennium Seed Bank was under construction at Wakehurst Place. It is costing a cool £80 million. £30 million of this has come from the Millennium Commission with national lottery money. But as with all such grants it had to be matched by similar funding from Kew. The Orange mobile phone company pitched in in the early days becoming Premier Sponsor with £2.5 million and shortly after the Wellcome Trust, a medical research foundation, came along with £9.2 million. Heady stuff. But Iain comments also that many, many small donors have put money in too, and spread the word, giving the project the oxygen of publicity and a ground swell of approval. Over half a million pounds has come from individual donors, Friends of Kew and many others. Again a significant element of the overall vision for the seed bank is its international connections. Countries taking part in the seed bank project can send teams to receive training not only in actual seed collection (for that is a science in itself), but also laboratory methods of storage should they want to set up seed banks in their own countries. In the mean time Wakehurst can offer banking facilities to a number of countries until such times as they build their own storage and the actual seed bank vaults, buried deep underground, form only one part of the facility with laboratories and accommodation for visiting scientists and researchers making up a significant part of the visible building complex. There is also be a full public display area so that all may come and see the on going work of this extraordinary ongoing project.

Maintaining biodiversity in this and other ways is one of Iain's deep personal concerns, combining as it does worldwide conservation, field work and - through identification for collection - the importance of taxonomy, his first passion.

"If you don't know what is out there you can't conserve it - except in the most general sense," he says. "For when you know what a plant is, what its family and genus are, you can usually foresee something of its value, economically, medically or environmentally. Thus the taxonomist of all people has a leading role to play in conservation and particularly the maintenance of biodiversity. For example a mere 5 per cent of the world's plant species have been properly tested for their medicinal properties. We have simply no idea what vital plants we may be losing every time another species becomes extinct. Maintaining biodiversity is important to give all life a broad base for existence, including our own. Life on earth depends on a full mix of species. We all inter-depend. That is how creation works."

Iain sees four overarching environmental problems which face the world, which have been clearly identified and now demand our attention: over-population, pollution, the greenhouse effect combined with the reduction of the ozone layer, and the loss of biodiversity.

Over-population he believes lies at the root of much of the environmental crisis as many communities, and especially multi-million population centres, struggle to support impossible ecologies. Even the seed bank reflects this - having to bank seeds from areas where human over exploitation use has made plants scarce. He remembers studying population numbers in Amazonian Indian villages many years ago:

"With up to about eighty people in a community most of the time there was harmony, but when numbers increased to about 120, fights broke out and disagreements occurred. Much beyond that and there would be a big fight and the village would break into two with one group going off to form another community elsewhere. The world is now populating exponentially (one billion in 1830 to five billion by 1987, six billion by 2000) and unless we do something I predict the same thing will happen on a global scale, except that, for us, there is no where else to go."

Pollution is in some ways the obverse of the population problem as over stretched communities turn to large scale industrial technology to meet the needs of overwhelming numbers. Acid rain, once a problem only for Scandanavia and Northern Europe, has become a worldwide phenomenon (a recent study in Eastern USA by the NYBG showed that the clouds there were more acid than lemon juice) and carbon dioxide, a product of combustion, continues to force the world average temperature inexorably higher through reinforcing the greenhouse effect of the atmosphere. It is now half a degree Celsius warmer everywhere than a century ago, but as half the carbon dioxide in the atmosphere has been put there in the last thirty years this rate of temperature increase is likely to speed up. Among the concerns is coastal flooding from melting ice caps. The ice melt-down has already started. And scientists are watching the glaciers and ice fields at the pole with increasing concern. Most of the major cities of the world are on the coast and flooding is well nigh impossible to stop.

There is also the effect of Methane (thirty times more potent than carbon dioxide in raising world temperature) which escapes from gas systems the world over, such as are used extensively in the former communist East and from landfill garbage sites in the West. CFC-12 gas (800 times more serious than methane) which also erodes the ozone layer - at first near the poles (in Argentina and Australia now skin-burn is a serious problem) but which will continue to creep up the globe unless CFCs are wholly banned. But many developing countries with first generation technology and facing great economic problems are not willing, or feel wholly unable, to do this.

But to stop or reverse these trends will mean a real sacrifice for everyone. Perhaps a loss of personal comforts to a level usually only expected of people in wartime. A significant reduction in the consumption of manufactured goods, reduced use of personal transport and fossil fuels and the spending of hard earned money on insulation and energy conservation measures. Though of course energy conservation actually saves money in the long run.

It is a lot to ask of those in the comfortable countries of the First World and of others who aspire to these comforts and are just getting their foot on the ladder. But Iain feels that if we don't ask it of ourselves now the planet will surely ask it of us later - in a cataclysmic and far less controllable way.

At the other, theoretical, end of scale Iain has also introduced the Kew Mission Statement and the Corporate Strategic Plan, Kew 2020, vision in here. Whilst a far cry from instituting conservation measures in the Palm House or speaking at a sponsors dinner or even plunging a thermometer into the flower head of *Victoria amazonica* in the field, Iain stoutly maintains that such statements and long term plans are just as important to the advancement of botany. And add continuity as he hands over his responsibilities.

The Kew Mission Statement is: 'To enable better management of the earth's environment by increasing knowledge and understanding of the plant kingdom - the basis of life on earth.'

As Iain moves on, to take up the reins as scientific advisor to the Eden Project that is building a 5 acre 150' tall rainforest in Cornwall, to continue to speak at conferences and dinners, seminars and symposiums, to visit lovingly his old Amazon haunts when he can, to continue his passion unabated, he feels that this mission statement would not be a bad summation of his life too. That, and service to God, to Iain the undisputed king of the kingdom of plants.

ACKNOWLEDGEMENTS
AND THANKS

There are so many people I must thank for help with my research. I trust I have remembered them all through all the many telephone calls (from transatlantic down to my own local village, near Malvern), the taped sessions, books, magazines and diaries I have reviewed and discussed, not to mention the tours around Kew and the New York Botanical Garden.

Joint thanks from Iain and myself to Anne, Iain's wife, and Sarah, his daughter. To Mickey Maroncelli, his indefatigable secretary from the NYBG, and Pauline Churcher and Eleanor Bunnell at Kew – whom I have worn out with requests for details of one kind or another. And broad thanks to the three institutions which feature so much: The New York Botanical Garden, Kew and the Instituto Nacional de Pesquisas da Amazônia.

Iain remembers with particular gratitude former directors of INPA: Djalma Batista, Paulo Machado and Warwick Kerr. And, while we are in the Amazon, field helpers José Ramos, Luiz Coelho, Dionisio Coelho and Osmarino P. Monteiro.

Also in Brazil: botanists Dr William Rodrigues, Dr Marlene F. da Silva, Dr Paulo B. Cavalcante and the late Dr João Murça Pires. Plus all the students of the 1973 INPA–FUA Masters Course.

Still in the New World: Dr Robert Goodland, now working for the World Bank, former NYBG Presidents Dr Howard Irwin Jr and the late Dr William C. Steere. And the late Dr Boris A. Krukoff, the 'Russian Bear', Iain's early Amazonian mentor.

Iain also feels a special debt to Dr Frank White, his DPhil supervisor, who sadly died in September 1994, and to H.C.W. (Bill) Wilson his housemaster who is hale and hearty in retirement near Malvern.

Iain also remembers with fondness and gratitude two of his own students who later became colleagues: Dr Eduardo Lleras and Dr Enrique Forero who aspired to great heights (Iain's job at the NYBG) and in the end got it.

I must thank so many at the New York Botanical Garden who kindly received Elizabeth, my wife, on her 'arctic' research trip to New York in the record low temperatures of January 1994. She didn't see too many plants under the ice and snow, though Bruce Riggs did his best on the tour (between digging the car out of snow drifts), but she did receive much warm hospitality from Dr Pat Holmgren who arranged everything, the NYBG President Gregory Long (who made doubly sure Elizabeth got safely back through the bad weather to the airport), Mickey Maroncelli and Rosemary Lawlor, Dr Scott Mori who presented his own 'Iain Prance slide show' for her, Dr Brian Boom and Pat's husband Noel, Dr Mike Balick who took over the Institute of Economic Botany from Iain and Andrew Henderson, a Britisher abroad, and not forgetting Ira Goode, the NYBG driver who met her from the plane.

I must also thank my additional UK helpers and advisors not so far included: Dr Nigel Hepper, formerly of Kew and a close Christian friend of Iain's, as is Sir Maurice Laing, Adine Keatley, Iain's sister, Martin Stanniforth who gave me a fascinating (particularly to me, a botanical ignoramus) tour of Kew and Mike Fay a very user-friendly plant scientist in 'microprop' who helped me understand a little of what went on there. Also Iain's former research assistant Kate Edwards in the Herbarium who explained to me the complexities of modern taxonomy (and dried Chrysobalanaceae) and Clare Dempsey who handled Pauline Churcher's overload on fraught occasions.

To complete the record of Iain's service at Kew my further research there and at Wakehurst Place requires some additional gratitude: First to my Kew publishing editor John Harris whose patience and gentle reminders have kept me to the task and to his department head Alyson Prior for energising this new printing in the first place. Thanks too to Suzy Dickerson, cloistered in the herbarium, for her detailed re-editing and to Dr Mark Chase in the Jodrell for his outline of molecular systematics, fitting me in hurriedly between parties of government ministers, such is the importance of his work. Also Roger Smith, Millenium Seed Bank director at Wakehurst, and Colette Grosse his resourceful PA, both sparing time for me whilst heavily engaged in the final stages of that impressive project. The indefatigable Christine Brandt at Kew's PR dept., and Dr Lindsey Davies, for the space to draft the final chapters. And finally, as ever, PA to the Director, par excellence, Eleanor Bunnell.

Clive Langmead
Malvern, January 1995

INDEX